SOYBEAN FOR HUMAN CONSUMPTION AND ANIMAL FEED

Edited by **Aleksandra Sudarić**

Soybean for Human Consumption and Animal Feed

http://dx.doi.org/10.5772/intechopen.73719
Edited by Aleksandra Sudarić

Contributors

Weiping Kong, Wenjiang Huang, Yinli Bi, Aleksandra Sudarić, Maja Matoša Kočar, Tomislav Duvnjak, Zvonimir Zdunić, Antonela Markulj Kulundžić, William Singer, Bo Zhang, M.A. Rouf Mian, Haibo Huang, Vikas Kumar, Marina Benito, Md. Sakhawat Hossain, Janice Ragaza, Howard Mason, Eric Joël Fofiri Nzossie, Christophe Bring

Notice

First published in London, United Kingdom, 2020 by IntechOpen
IntechOpen is the global imprint of INTECHOPEN LIMITED, registered in England and Wales, registration number: 11086078, 5 Princes Gate Court, London, SW7 2QJ, United Kingdom
Printed in Croatia

British Library Cataloguing-in-Publication Data
A catalogue record for this book is available from the British Library

Additional hard and PDF copies can be obtained from orders@intechopen.com

Soybean for Human Consumption and Animal Feed, Edited by Aleksandra Sudarić
p. cm.
Print ISBN 978-1-83881-018-4
Online ISBN 978-1-83881-019-1
eBook (PDF) ISBN 978-1-83881-020-7

For EU product safety concerns:
IN TECH d.o.o., Prolaz Marije Krucifikse Kozulić 3, 51000 Rijeka, Croatia,
info@intechopen.com or visit our website at intechopen.com.

Meet the editor

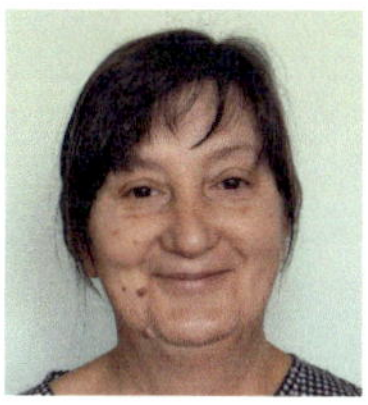

Dr. Aleksandra Sudarić was born in Osijek, Croatia. She received an MSc in Plant Genetics and Breeding in 1996 from the University of Zagreb and a PhD in Plant Breeding in 1999 from the University of Osijek. Dr. Sudarić is a senior research scientist at the Agricultural Institute Osijek (Croatia), Department for Breeding and Genetics of Industrial Plants where she studies the breeding and seed production of soybean. She is presently coordinator of the soybean breeding program at the Institute. In addition, she completed the training for application of DNA-based marker technology at the University of Guelph in Guelph (Canada). Her main research interests are in the genetic improvement of quantity and quality characteristics of soybean and soybean adaptation to mega-environments. Dr. Sudarić has authored more than 100 papers (scientific, professional, and popular), several book chapters, books, and manuscripts about soybean as well as developed twenty-six soybean varieties.

Contents

Preface

As food science in general has continued to evolve, and as the scientific evidence on the health benefits of particular food attributes has become more familiar to consumer expectations and needs, the transfer of new research findings into applications has become more challenging for food producers. Taking into consideration physiological functions of some food constituents, as well as their nutritional implications, it is necessary to point towards better utilization of health-promoting ingredients. Soybean (*Glycine max* (L.) Merr.) is the main oilseed crop worldwide, a staple crop for protein-rich food and feed as well as a significant source of nutraceutical compounds with many different medical benefits. *Soybean for Human Consumption and Animal Feed* provides an overview of the present state of the research of soybean nutritional attributes as well as science-based approaches defining the future of soybean for human consumption and animal feed.

The introductory chapter of this book presents a comprehensive picture of the potential of soybean for food, feed, and the pharmaceutical industry with an emphasis on the importance of genetic improvement of soybean germplasm in enhancing healthy and safe properties of final soy products.

Soybean is an important source of protein and amino acids for humans and livestock because of its well-balanced amino acid profile. Breeding efforts, genetic engineering efforts, agronomic properties, and processing methods to improve the nutritional quality of soybean in regard to amino acid levels as well as evaluation methods for amino acid quantification are discussed in Chapter 2 – "Soybean Amino Acids in Health, Genetics and Evaluation", written by W.S. Singer et al.

Chapter 3, written by A. Sudarić et al., examines the most important traits of soybean seed quality: protein content and amino acid composition, oil content and fatty acid composition, soluble sugars, and isoflavones. , written by A. Sudarić et al. this chapter gives an overview of European soybean seed quality research with an emphasis on the importance of genetic enhancement of soybean seed quality in improving the added-value properties of final soybean products.

The scientific and technological developments in areas of soybean utilization have increased exponentially. With numerous benefits, soya is recognized as a dietary allergen, containing a number of identified allergenic proteins. Chapter 4 by H. J. Mason refers the current knowledge of the respiratory health risks from exposure to dust from soybean and its products. The chapter also highlights other potential health hazards associated with soybean, such as endotoxin and fungal contamination.

Soybean protein is the major source of protein as fishmeal replacement in fish feed because of its worldwide availability and low price. Chapter 5, written by V. Kumar et al., focuses on the current knowledge about the potential impacts of different soybean protein inclusion in aqua feed on gut health condition of fish with special emphasis on gut morphology, soybean meal-induced enteritis, gut-associated immunity, and gut microbiota.

Chapter 6 reviews newer technologies in the realm of soybean monitoring in order to increase soybean tolerance to abiotic stress, primarily drought. W. Kong et al. examine possibilities for determining and selecting genotypes tolerant to abiotic stress. The interdisciplinary topic of the chapter introduces modern soybean breeding strategies, a crucial part in increasing soya food production and enhancing food security and sustainability.

Chapter 7 shows socioeconomic aspects of soybean production in the African rural areas (Sudano-Sahelian region). E.J. Fofiri Nzossié and C. Bring analyze the sustainability challenges of the product and market dynamics built around soybeans, speculation subject to strong requirements of international competitiveness and presenting at the same time a strong potential for improving farmers' incomes.

The references at the end of each chapter provide a starting point to acquire a deeper knowledge of the state of the art of the topic. This book is an interesting and helpful resource for specialists and students in the field of food production, primarily soya foods.

The book is a result of efforts by many scientists from different countries. I would like to acknowledge each of the authors who devoted much time and effort in delivering their chapters. We hope that this book will contribute to improving information about the high nutritional value of soybean and its potential in human and animal diets and health and as quality stock for pharmaceutical and functional food industries.

Dr. Aleksandra Sudarić
Agricultural Institute Osijek
Osijek, Croatia

Centre of Excellence for Biodiversity and Molecular Plant Breeding, Faculty of Agriculture,
University of Zagreb
Zagreb, Croatia

Chapter 1

Introductory Chapter: Soybean - Quality and Utilization

Aleksandra Sudarić

Additional information is available at the end of the chapter

http://dx.doi.org/10.5772/intechopen.93942

1. Introduction

Customer requirements regarding safe product of high quality and nutritional value are increasing every day. These requirements are the results of today's life style, market globalization, food born outbreaks, etc. In order to fulfill those requirements, institutions in food chain (from field to table) are reaching for tools to enhance food quality as well as to ensure the safety of nutritious, functional, healthy and quality foods. As the food science in general has continued to evolve and as the scientific evidence on the health benefits of particular food attributes has become more familiar to consumer expectations and needs, the transfer of new research findings into applications has become more challenging for food producers. Taking into consideration physiological functions of some food constituents, as well as their nutritional implications, it is necessary to point towards better utilization of health promoting ingredients (natural antioxidants, phytochemicals, vitamins, fortification ingredients) The final aim of food scientist's activity in the future, is to connect further food product and process development with consumer physiological needs, based on research results obtained in respective fundamental disciplines [1].

Soybean (*Glycine max* (L.) Merr.) is a species of legume native of the Eastern Asia, widely grown for its edible bean, which has numerous uses. For many centuries, it was the main source of food for the people of the Far East (China, Japan, India). The ability of soybean for the food and non-food uses has been recognized for many years. Among available protein sources of plant origin, soybean protein is the closest to the optimum dietary essential amino profiles for human and animal nutrition. Therefore, soybean production, consumer acceptance and consumption in non-traditional regions of the world are on the rise. Increasing world population, constant need for animal feed and numerous different soybean products contribute to the great demands for soybean seed in world market. Nowadays, soybean is the leading oil and protein crop, whose seed is used as a source of edible oil and protein for both human and livestock consumption and for various industrial purposes. Due to consumer's large interest in healthy food, soybean is grown more and more frequently under the organic system in which it produces satisfactory yield [2, 3].

2. Seed quality

The importance of soybean stems from the quality of its seed [4–6]. It contains about 35-50% proteins and 18-24% oils, depending of the variety and growing conditions. Soybean has the highest crude protein content and a balanced amino acid profile. Hence, soybeans have become a top choice for protein and as such, they are strategically traded around the world. The main protease inhibitors in soybean – Kunitz trypsin inhibitor and the Bowman-Birk inhibitor constitute the main anti-nutritional factors of soybean. Trypsin inhibitors contribute to reduce digestibility of seed protein. Nutritional and functional properties of the soya oils are determined by their fatty acid composition, the distribution pattern of the fatty acids within the triacylglycerols (TAGs) and the total content and composition of natural antioxidants [7]. Soybean is considered to be the most abundant natural source of isoflavones in the human and animal diet [8]. Isoflavones are the main components of flavonoids and the most common form of phytoestrogens [6]. Totally dry soybean seed contains 33% DW of carbohydrates on average, of which 16.6% DW is soluble sugars [9]. The significance of the soluble sugar profile is in its effect on quality, digestibility and nutritional value of soybean for food and feed. Soybean seed also contains vitamins (A, B, D and E), minerals (calcium, iron, and potassium) and phospholipids, such as lecithin.

3. Utilization

Soybean is used both for oil production and protein processing [5, 10]. Most of the produced soybean in the world is used for animal nutrition. Soybean meal is the highest quality plant protein nutrient that can serve as the basic source of protein for animal, especially all categories of pigs and poultry, then fish, fattening cattle, and is important in the diet of high-milking cows and sports horses. The use of raw soybean in the diet does not give satisfactory results due to trypsin inhibitors. Before using as food or animal feed, soybean seed must be heat-treated to break down trypsin inhibitor's activity. Processed soybean foods include various products such as tofu (bean curd), yuba (soy protein extracts from soy milk), kinako (roasted soy powder), nimane (cooked beans), bean sprouts, and soymilk as non-fermented food products and soy sauce, miso, and natto as fermented food products. Among these soy-based food products, tofu, soy sauce, miso, bean sprouts and natto are predominant. Soybean oil has been traditionally appreciated as a high quality commodity in the world oil market and due to widely has used in many food and non-food applications [10, 11]. Soybean oil is increasingly used in the food industry – for frying or baking foods, as a vegetable oil (table oil), as an ingredient in foods like salad dressings and margarins. The raw material is in the industry - soaps, creams, detergents, paints and varnishes, plastics, cleaners, medical preparations, ink. Lecitin from soybean oil has applications in the bakery, confectionary, pharmaceutical, textile and chemical industries. Soybean is also used to produce biodiesel. Biodisel *"methyl soyate"* is a renewable substitute for petroleum diesel with reduced greenhouse gas emissions. Results of numerous medical studies have indicated on the important role of soybean in the prevention and curing of chronic diseases. Healthy aspects of soyfoods go beyond the oil and protein and include minor compounds with nutraceutical properties [12, 13].

4. Perspectives

Soybean breeding has undoubtedly played a key role in production increases. The genetic improvement of soybean germplasm, based on conventional breeding strategies, contributes to advances in production and food processing industry by developing high-yielding and high-quality soybean varieties, hereby enhancing value-added, healthy and safe properties of final soy products. Recent advances in biotechnology, in particular the development of improved molecular marker technology, have made possible the genetic dissection and characterization of many quantitatively inherited seed quality traits in soybean [5, 14–17]. Therefore, genetic improvement of varieties could be characterized as an integral part of sustainable food production. The value of soya in livestock and human nutrition and industry has not yet been fully exploited. The processing methods in the animal feed industry and food, chemical, pharmaceutical and other industries are continuously improving.

Scientists look for opportunities to leverage research capabilities and commercialize promising new technologies by cooperating and collaborating with public and private sector interests. It is likely that these efforts will maintain a strong focus on soybean as a source of food and industrial products well into the future.

Author details

Aleksandra Sudarić[1,2*]

*Address all correspondence to: aleksandra.sudaric@poljinos.hr

1 Agricultural Institute Osijek, Osijek, Republic of Croatia

2 Faculty of Agriculture, Centre of Excellence for Biodiversity and Molecular Plant Breeding, University of Zagreb, Zagreb, Republic of Croatia

References

[1] Aladjadjiyan A. Physical Factors for Plant Growth Stimulation Improve Food Quality. In: Aladjadjiyan, A., editor. Food Production - Approaches, Challenges and Tasks. Rijeka (Croatia): InTech; 2012. p. 145-168.

[2] McBride WD, Greene C. The profitability of organic soybean production. Renewable Agriculture and Food Systems. 2009, 24 (4): 276-284.

[3] Balešević-Tubić S, Đukić V. Organic soybean production. Legume Perspectives. 2013; 1:48-49.

[4] Wilson RF. Seed Composion. In: Boerma, H.R. and Specht, J.E., editors. Soybeans: Improvement, Production and Uses. Medison (USA): ASA, CSSA, SSSA, 2004. p. 621-677.

[5] Cober ER, Cianzio SR, Pantalone VR. Soybean. In: Vollmann, J. and Rajcan I. editors. Oil Crops. Springer-Verlag New York Inc, 2009. p. 57-90.

[6] Kim EH, Ro HM, Kim SL, Kim HS, Chung IM. Analysis of isoflavone, phenolic, soyasapogenol, and tocopherol compounds in soybean (*Glycine max* (L.) Merrill) germplasms of different seed weights and origins. Journal of Agricultural and Food Chemistry. 2012; 60:6045-6055.

[7] Velasco L, Fernandez-Martinez JM. Breeding Oilseed Crops for Improved Oil Quality. In: Basra, A.S. and Randhawa, L.S. editors. Quality Improvement in Field Crops. Oxford, Food Product Press, 2002. p. 309-344.

[8] Bhagwat S. Haytowitz DB, Holden JM. USDA Database for the Isoflavone Content of Selected Foods, Release 2.0. U.S. Department of Agriculture, Agricultural Research Service, NutrientData Laboratory. http://www.ars.usda.gov/nutrientdata/isoflav

[9] Hymowitz T, Collins FI. Variability of sugar content in seed of *Glycine max* (L.) Merrill. and G.soja Sieb. and Zucc. Agronomy Journal. 1974;66 (2):239-240.

[10] Lusas EW. Soybean Processing and Utilization. In: Boerma, H.R. and Specht, J.E., editors. Soybeans: Improvement, Production and Uses. Medison (USA): ASA, CSSA, SSSA, 2004. p. 949-1045.

[11] Rebetzke GJ, Burton JW, Carter Jr EC, Wilson RF. Changes in Agronomic and Seed Characteristics with Selection for Reduced Palmitic Acid Content in Soybean. Crop Science. 1998, 38(2):297-302.

[12] Rajcan I, Hou G, Weir AD. Advances in Breeding of Seed Quality Traits in Soybean. In: Kang MS editor. Genetic and production Innovations in Field Crop Technology: New Development in Theory and Practice. Food Product Press, 2005. p. 145-174.

[13] Jayachandran M., Xu B. An insight into the health benefits of fermented soy products. Food Chemistry. 2015, 271:362-371.

[14] Hemingway J, Eskandari M, Rajcan I. Genetic and Environmental Effects on Fatty Acid Composition in Soybeans with Potential Use in Automotive Industry. Crop Science. 2015, 51:1-11.

[15] Shaw EJ, Rajcan I. Molecular Mapping of Soybean Seed Tocopherols in the cross OAC Bayfield x OAC Shire. Plant Breeding. 2017, 136:83-93.

[16] Watanabe D, Lošak T, Vollmann J. From proteomics to ionomics: Soybean genetic improvement for better food safety. Genetika. 2018, 50:333-350.

[17] Vollmann J, Sagara T, Pachner M, Schmidt A, Mayer HK. The polyamine spermidine as a health component in soy-foods: From soybean breeding to food production. International Legume Society and Institute of Plant Genetics, Polish Acdemy of Sciences editors. Third International Legume Society Conference Legumes for human health and planet health. 21-24 May 2019, Poznan, Poland Book of Abstracts, ISBN 978-83-950380-1-3. p. 163.

Chapter 2

Soybean Amino Acids in Health, Genetics, and Evaluation

William Monte Singer, Bo Zhang,
M.A. Rouf Mian and Haibo Huang

Additional information is available at the end of the chapter

http://dx.doi.org/10.5772/intechopen.89497

Abstract

Soybean is an important source of protein and amino acids for humans and livestock because of its well-balanced amino acid profile. This chapter outlines the strengths and weaknesses of soybean as a complete amino acid source as well as the relative importance of individual amino acids. Special attention is paid to the sulfur-containing amino acids, methionine and cysteine. Breeding and genetic engineering efforts are summarized to highlight previous accomplishments in amino acid improvement and potential avenues for future research. Agronomic properties and processing methods that affect amino acid levels in soybean food and feed are also explained. A brief introduction into current amino acid evaluation techniques is provided. By understanding the complexities of amino acids in soybean, protein quality for humans and livestock can be maximized.

Keywords: essential and nonessential amino acids, soybean meal, methionine, cysteine, sulfur-containing amino acids

1. Introduction

Soybean is one of the world's most economically and nutritionally important crops. In 2018, soybean were 61% of international oilseed production with 397.9 tons harvested worldwide [1]. The United States and Brazil were the largest producers at 4545 and 4299 million bushels, respectively, with China being the largest importer of U.S. whole soybeans valued over $3 billion U.S. [2, 3]. Soybean products, namely meal and oil, are popular in a myriad of industries for their versatility and utility. Soybean oil provides the most versatility with uses in fuel, solvents, candles, cosmetics, construction, and foam. However, soybean meal is the driving factor for 70% of the plant's value with 97% of all U.S. soybean meal being used for animal feed [4]. As such, an enormous portion of soybean's importance lies with its nutritional capabilities for livestock and humans.

Nutritionally speaking, soybeans are a highly valued protein source. Proteins are a crucial macromolecule needed in the diets of human and livestock. However, the true significance of soybean protein is due to its well-balanced amino acid profile that aligns with dietary needs of humans and animals [5]. Amino acids are the functional subunits of proteins that, when linked together in different orders, generate the variety of proteins critical to life. Amino acids are also important intermediates for many biosynthesis pathways [6]. Deficiencies in single or multiple amino acids can negatively impact an individual's growth and development [7, 8]. Intriguingly, an excess of certain amino acids has also been shown to worsen feed intake, nitrogen efficiency, and growth rate in livestock [9–11]. The importance of amino acid levels on human health has also been well documented [12, 13].

Amino acids are characterized by having amine ($-NH_2$) and carboxyl ($-COOH$) functional groups as well as a "R-group" that is unique to each amino acid [14]. Amino acids are abundant in both proteinogenic (protein-incorporated) and non-proteinogenic forms [15]. The 20 common, proteinogenic amino acids are generally the focus of research in soybeans as they are the defining nutritive feature. Of those twenty, nine amino acids are essential for humans to consume. Livestock usually require these same amino acids from feed and might require others because of their biological systems [16]. Soybeans contain some level of all nine essential amino acids which creates a suitable nutritional foundation for livestock feed and human food.

2. Essential amino acids

Essential amino acids are ones that living organisms are unable to biosynthesize themselves and must obtain from their food source [7, 12, 16]. Therefore, in this term, "essential" refers to the amino acid requirements in dietary ingredients. The nine standard essential amino acids for humans present in soybean are: histidine, isoleucine, leucine, lysine, methionine, phenylalanine, threonine, tryptophan, and valine [17]. Arginine is regularly considered an essential amino acid for fish, poultry and sometimes swine due to absent or deficient urea cycles [9, 16]. Poultry and reptiles also require dietary glycine because of differing waste excretion pathways [16]. While crude protein content is normally recognized as the driving nutritional factor for soybean meal, these essential amino acids provide true utility.

It has long been recommended that protein quality is based upon essential amino acid content. However, for many reasons, animal feed and human food markets have only recently begun assessing accordingly [5, 18, 19]. Equipment required for accurate amino acid measurement and the diversity of markets for amino acids makes it difficult for supply chain evaluators like elevator operators to appraise amino acid content on site. To some degree, the well-balanced soybean amino acid profile also devalues the need to measure individual amino acid levels. Since all essential amino acids are present, less attention is paid to deficient amino acids such as methionine and tryptophan [17, 20].

Deficiencies in soybean's essential amino acid profile has led to a large section of the livestock industry focusing on feed mixing and supplementation. Rationing with other feed sources such as cereal grain and synthetic amino acid augmentation can effectively resolve the issue. Although, this comes with economic and environmental problems. Supplementing amino acids adds costs to farmers. For example, the average cost for amino acids supplementation for dairy farmers is 20 cents per head per day [21]. Maximizing crude protein for a growth limiting factor also negatively impacts livestock nitrogen-use-efficiency and environmental nitrogen outputs [8, 22]. Synthetic amino acid production can produce hazardous environmental waste and synthetic methionine, the most limiting soybean amino acid for poultry, has also been banned for organic poultry production [20, 23]. Movement towards sustainable agriculture will pressure the feed industry to alter how soybean meal is enhanced for essential amino acid livestock maximization. Furthermore, the increasing popularity of meat-less diets in humans will create new markets for soybean's well-balanced amino acid profile.

3. Nonessential amino acids

Nonessential amino acids should not be misconstrued as unimportant amino acids. Of the 20 proteinogenic amino acids, those considered nonessential are still necessary for living organisms. Healthy organisms are able to biosynthesize them and are not required from food and feed consumption. The 20 standard nonessential amino acids for humans found in soybean are: arginine, alanine, asparagine, aspartic acid, cysteine, glutamine, glutamic acid, glycine, proline, serine, and tyrosine [24–26]. As previously mentioned, the necessity of amino acids such as arginine and glycine can differ amongst species. Some nonessential amino acids are also affected by the presence and amounts of essential amino acids.

Cysteine not provided from food consumption is directly biosynthesized from methionine via trans-sulfuration [16, 27–30]. Consequently, if cysteine is not provided in the diet, then enough methionine must be provided to compensate for both amino acid needs. For that reason, feed research for poultry occasionally measures methionine and cysteine jointly [10, 31, 32]. Tyrosine is also directly formed from phenylalanine via hydroxylation [16, 29, 33]. Other amino acids like arginine, glycine, and proline can be required from the diet when animals are young, old, sick, or otherwise deficient in body protein regulation. As human and livestock diets become more sustainably plant-based, it will become more important to evaluate nonessential amino acids, specifically the ones immediately affected by essential amino acids.

4. Proteinogenic sulfur-containing amino acids

The two proteinogenic sulfur-containing amino acids, methionine and cysteine, are critical to evaluating soybean meal as food and feed. While present in soybean, methionine and cysteine levels are both inadequate for consumer needs [8, 20, 24]. Similar to research determinations, the nutritional requirements for methionine and cysteine intake are often grouped together as overall sulfur consumption from protein. Adult humans are recommended to intake 910–1120 mg methionine and cysteine (based on body weight) per day [13]. For livestock, sulfur amino acids recommendations can vary based on species, age, end-use, and diet formulation. The importance of dietary sulfur amino acids for livestock is greatly emphasized in literature, especially with soybean meal as base feed [8, 20, 34, 35].

Methionine and cysteine are vital to biological functions because of the sulfur contained in their R-groups and versatility in macromolecule synthesis. Methionine is well-known for being the typical initiating amino acid for protein synthesis and has hydrophobic properties when incorporated into proteins [36]. These hydrophobic properties usually result in methionine incorporation within the core of proteins. However, certain proteins have surface-exposed methionine susceptible to oxidation that is associated with age-related disease [30, 37, 38]. S-adenosylmethionine, a methionine metabolism intermediate, is widely-used with functions in methylation as well as amine, methylene, and sulfur atom donation [30, 39, 40]. Methionine is especially important for poultry production as birds have exceedingly high sulfur amino acid requirements, and low methionine levels can negatively affect growth rate, carcass yield, fat content, and disease immunity [41–43]. Cysteine's ability to form disulfide bonds makes it incredibly important to tertiary protein structure and can occur with and without enzyme interactions [30, 44]. Cysteine is involved with keratin and feather production in poultry and deficiencies have been correlated to poor breast muscle development [45, 46]. Swine also need higher amounts of cysteine as they age to compensate for body maintenance [29, 47].

5. Breeding efforts

Soybean has an extensive cultivated history dating back thousands of years to its country of origin, China. *Glycine max*, the contemporary species of soybean, was domesticated from the wild species *Glycine soja* and has been continually been improved through selective and molecular breeding [48–50]. Once harvesting traits such as seed shatter and lodging were improved in the late 1930s to make soybean a competitive row crop, other cultivar improvements became a valuable research goal [49]. Current breeding programs tend to focus on traits such as yield, disease resistance, abiotic stress tolerance, and seed composition. Seed composition improvements include protein and oil content, fatty acid levels, anti-nutritional factors, isoflavones, and amino acids profiles. Before 1972, there had been zero reported research for improvement of soybean amino acid profiles, rather with emphasis on overall protein content [51]. Modern breeders are also inclined to concentrate efforts on protein content and consider amino acid levels an afterthought [52]. TN04-5321 is the only released germplasm in the United States that maintains yield and protein content while improving amino acid balance by increasing methionine and cysteine to levels recommended livestock needs [53].

The major soybean storage proteins are 11S (glycinin) and 7S (conglycinin) and provide the bulk of amino acids with limited non-proteinogenic amino acids in seed [54–56]. By increasing 11S and 7S quantity, more protein can ultimately be present in food and feed. Overemphasis on crude protein content can have negative ramifications on overall protein quality, specifically deficient amino acids. While an increase in protein content would theoretically entail an increase in amino acids including methionine and cysteine [57], the opposite effect has been more notable [58, 59]. Molecular breeding techniques have recently improved the understanding of amino acid genomic regulation. Multiple quantitative trait loci (QTL) studies have been performed to identify genomic regions that control amino acids levels in soybean seed [60–64]. A myriad of QTL's were found to create amino acid phenotypic variation. Individual amino acids had reoccurring or proximal QTL's discovered such as Satt 518 [60], ss107913002 [61], and BARC-048619 [62] for glycine and threonine. QTL's for methionine and cysteine were also discovered which could lead to valuable improvements for soybean livestock feed [60–64]. Other genomic studies such as genome-wide association studies (GWAS) and genetic diversity analyses would further improve genetic understanding.

6. Genetic engineering

Genetic engineering experiments such as genetically modified organisms (GMOs) and gene editing are also promising avenues for improving amino acid profiles of soybeans. Compared to conventionally bred varieties, transgenic soybeans face additional adversity from registration requirements and public opinion. Transgenic efforts generally have one of three targets: magnifying biosynthesis genes, adjusting biosynthesis regulation, and modifying storage proteins. The earliest example would be a Brazil nut gene transfer in 1992. This successfully increased protein content and methionine biosynthesis, however a major food allergen was also transferred making commercialization impossible [65]. Expressing zein proteins from corn has also been shown to increase sulfur-containing amino acids levels in soybean [66, 67]. Altering biosynthesis feedback regulation amplified both non-proteinogenic and proteogenic lysine by circumventing normal enzymatic pathways [68]. Tryptophan in soybean also exhibited increased non-proteinogenic levels when a feedback-insensitive enzyme was transferred [25]. While soybean is not deficient in lysine or tryptophan, corn is deficient in both. By increasing lysine and tryptophan concentrations, soybean becomes an even more useful feed additive to corn rations. Even with limited research on modifying overall amino acids profiles in soybeans, modifying 11S and 7S storage proteins ratios [69] or silencing their expression entirely [70] has displayed increased amino acids levels. Similarly, a study using irradiated mutant soybeans lacking storage proteins as breeding parents demonstrated increased non-proteinogenic amino acids contents [54]. In addition, further research should be conducted to determine the bioavailability and digestibility of increased non-proteinogenic amino acids in soybean.

7. Agronomic relations

Amino acids concentrations in soybean are not only affected by their genetic potential. Agronomic properties greatly impact the final levels of amino acids. Agronomy encompasses all aspects of crop production including environmental effects, climatic variables, and abiotic factors. Perhaps the most considered agronomic factor is soil nutrient availability. Insufficient soil nutrient levels of nitrogen, potassium, phosphorous, sulfur, calcium, and magnesium create poor amino acid profiles in soybean plants [71]. Increased phosphorous rates have been shown to increase the percentage of methionine and tryptophan in seed but had no effect on protein content percentage [72]. Applications of sulfur, phosphorous, and nitrogen (individual and combined) produced a variety of different methionine and cysteine seed concentrations [73]. Sulfur deficiencies were also shown to inhibit the production of 11S proteins while almost eliminating methionine and cysteine in 7S proteins [74]. It is becoming more popular to also apply biological substances such as amino acids to plants through foliar and seed application. Amino acid uptake by soybean and wheat have been proven, and improved soybean growth rates and antioxidant effects have also occurred [75, 76]. Further research should be conducted to determine if biofortification solutions are possible through amino acid application.

Amino acid variation has also been shown to occur across environments [24, 59]. Specific correlations have emerged in response to temperature, solar radiation, and rainfall. One study shows that increased temperature leads to increased concentrations of all proteinogenic amino acids [77], while another concludes that only methionine and cysteine increase alongside temperature [78]. Increased solar radiation and greater available water appeared to have a negative relationship with amino acid content [77]. These favorable conditions would increase yield which has been shown to have a negative correlation with overall protein content [57]. The multitude of agronomic factors that affect amino acid profiles in soybean make it exceedingly important to compensate for variables when researching.

8. Processing impacts

The diversity in food, feed, and industrial used for soybean require the whole seed or seed components to be processed. Processing can affect the nutritional value of soybean protein and presence of amino acids in food and feed. Processing procedures can either separate seed components for different purposes or convert the entire seed into a product (usually human food). Some human soy foods such as edamame and soybean sprouts need little to no processing. Others including soymilk, tofu, natto, and soy sauce involve more processing. Soymilk and tofu processing are interconnected. Soymilk is a water-extract of whole or crushed soybeans that is coagulated and pressed into tofu [48]. While not all seed proteins convert into protein in tofu, 11S/7S storage protein ratios have been shown to be both positively and negatively correlated with tofu hardness [79, 80]. Natto is a soy food created by fermenting whole soybeans with *Bacillus subtilis*. Fermentation time affects final amino acid concentrations, and proper fermentation length could potentially increase nutritional values [81]. Soy sauce is produced by traditional and commercial methods, but both are based around whole seed or meal fermentation with *Aspergillus* sp. However, commercial methods have a lower amino acid to nitrogen ratio [48].

Soybean meal processing also impacts the level of amino acids in livestock feed. The first step in soybean meal processing is essentially separating protein from oil. A variety of methods exist including solvent extraction, screw pressing, and extruding [48, 82, 83]. All three processes have three final products: oil, meal (usually toasted to lessen anti-nutritional factors), and hulls. Over processing of solvent extracted soymeal has been shown to decrease lysine, cysteine, and arginine levels [84, 85]. Protein solubility and dispersibility measurements may be a useful indicator of over processing [86, 87]. Soybean hulls are sometimes added to livestock feed for additional fiber, however an increase in hull/meal ratios decrease the digestibility of amino acids [88]. While soybean is renowned for its protein and amino acid content, actual nutritional values can be decreased through certain processing methods.

9. Evaluation methods

All previously mentioned aspects of soybean production in regard to amino acid levels and human and animal nutrition depend on a single common denominator: amino acid quantification. Amino acids must be reliably, effectively, and accurately identified, measured and evaluated. A Google Scholar search of "amino acid analysis" will display over 1 million results. Several reviews have been published regarding the development of amino acid analysis [89–91]. In general, contemporary analysis of amino acids from any source will be performed by chromatography or near-infrared reflectance spectroscopy. Chromatography is the common method with specific techniques including ion exchange chromatography (IEC), high-performance liquid chromatography (HPLC), and gas chromatography (GC). HPLC is the more validated method for soybean amino acid analysis. It is more efficient than IEC, and it does not require the transformation into volatiles like GC [91–94]. Near-infrared reflectance spectroscopy (NIRS) is a more recent addition to amino acid analysis, and it has the potential to drastically improve the efficiency in soybean feed evaluation [19]. The inability to actually measure amino acid levels is main hindrance for NIRS amino acid analysis. NIRS methods must be developed from a calibration set of raw data (often from HPLC) [95–97]. Nonetheless, efficiency improvements should persuade researchers to continually explore future NIRS amino acid analysis applications.

10. Conclusion

Soybean is a valuable source of protein and amino acids for humans and livestock. Soybean's well-balanced amino acid profile provides all essential amino acids as well as most nonessential. However, there is much room for nutritional improvement. Proteinogenic sulfur-containing amino acids, methionine and cysteine, are deficient in soybean and are especially needed in livestock rations. Increased levels of these amino acids would augment soybean meal value and lessen the need for synthetic amino acid supplements. Breeding efforts have made little progress in adjusting amino acid profiles thus far, however significant developments in understanding genomic control regions promise future success. Genetic engineering efforts have shown promising amino acid improvements, but regulations and public opinion made commercialization difficult. New gene-editing technology could be the key to unlocking true nutritional improvement.

Agronomic properties and processing methods both impact the final quantities of amino acids available to humans and livestock. Understanding these impacts are essential to improve the nutritional quality of soybeans. Amino acid evaluation through HPLC provides reliable and efficient quantification, yet even quicker measurements are possible through NIRS. As the world's population continues to grow, soybeans will be essential to both human and livestock for amino acid requirements. Wholesome approaches that understand the complexities of amino acids in soybean will be required to maximize overall success and feed the world with balance soy proteins.

Acknowledgements

The authors would like to acknowledge Virginia Polytechnic Institute and State University's Open Access Subvention Fund for funding the publishing fee.

Conflict of interest

The authors declare there is no conflict of interest.

Author details

William Monte Singer[1*], Bo Zhang[1], M.A. Rouf Mian[2] and Haibo Huang[3]

*Address all correspondence to: wilmsing@vt.edu

1 School of Plant and Environmental Sciences, Virginia Polytechnic and State Institution, Blacksburg, Virginia, USA

2 United States Department of Agriculture, Agricultural Research Service, Raleigh, North Carolina

3 Department of Food Science and Technology, Virginia Polytechnic Institute and State University, Blacksburg, Virginia, USA

References

[1] International: World Oilseed Production. 2019. Available from: http://soystats.com/international-world-oilseed-production/

[2] International: World Soybean Production. 2019. Available from: http://soystats.com/international-world-soybean-production/

[3] U.S. Exports: Soy Exports by Customer. 2019. Available from: http://soystats.com/u-s-exports-soy-exports-by-customersda/

[4] Soybean Meal. United Soybean Board. 2019. Available from: https://unitedsoybean.org/topics/soybean-meal/

[5] Osborne TB, Mendel LB. Amino-acids in nutrition and growth. 1914. Journal of the American College of Nutrition. 1993;**12**(5):484-485

[6] Herrmann K, Somerville RL. Amino Acids: Biosynthesis and Genetic Regulation. Reading, Massachusetts: Addison-Wesley Publishing Company, Inc; 1983

[7] Wade C. Carlson Wade's Amino Acids Book. New Canaan, Connecticut: Keats Publishing Inc.; 1985

[8] Berry TH, Becker DE, Rasmussen OG, Jensen AH, Norton HW. The limiting amino acids in soybean protein. Journal of Animal Science. 1962;**21**(3):558-561

[9] Boisen S, Hvelplund T, Weisbjerg MR. Ideal amino acid profiles as a basis for feed protein evaluation. Livestock Production Science. 2000;**64**(2):239-251

[10] Han Y, Suzuki H, Parsons CM, Baker DH. Amino acid fortification of a low-protein corn and soybean meal diet for chicks. Poultry Science. 1992;**71**(7):1168-1178

[11] Waldroup PW, Mitchell RJ, Payne JR, Hazen KR. Performance of chicks fed diets formulated to minimize excess levels of essential amino acids. Poultry Science. 1976;**55**(1):243-253

[12] D'Mello JPF, editor. Amino Acids in Human Nutrition and Health. Wallingford, United Kingdom: CAB International; 2012

[13] Jez JM, Fukagawa NK. Plant sulphur compounds and human health. In: Sulfur: A Missing Link between Soils, Crops, and Nutrition. American Society of Agronomy, Crop Science Society of America, Soil Science Society of America; 2008. pp. 281-291

[14] Weaver R. Molecular Biology. 4th ed. New York, NY: McGraw-Hill; 2008

[15] Wagner I, Musso H. New naturally occurring amino acids. Angewandte Chemie (International Ed. in English). 1983;**22**(11):816-828

[16] Buttery PJ, D'Mello JPF. Amino acid metabolism in farm animals: An overview. In: Amino Acids in Farm Animal Nutrition. Wallingford, United Kingdom: CAB International; 1994. pp. 1-10

[17] Kuiken KA, Lyman M. Essential amino acid composition of soy bean meals prepared from twenty strains of soy beans. Journal of Biological Chemistry. 1949;**177**:29-36

[18] Pfarr MD, Kazula MJ, Miller-Garvin JE, Naeve SL. Amino acid balance is affected by protein concentration in soybean. Crop Science. 2018;**58**(5):2050-2062

[19] Measuring Beyond the Bushel. United Soybean Board. 2019. Available from: https://unitedsoybean.org/article/measuring-beyond-the-bushel-1/

[20] Fernandez SR, Aoyagi S, Han Y, Parsons CM, Baker DH. Limiting order of amino acids in corn and soybean meal for growth of the chick. Poultry Science. 1994;**73**(12):1887-1896

[21] Control Feed Costs with Amino Acids. Drovers. 2019. Available from: https://www.drovers.com/article/control-feed-costs-amino-acids

[22] Managing Nutrients and Pathogens from Animal Agriculture. Natural Resource, Agriculture, and Engineering Service. Ithaca, NY; 2000

[23] Willke T. Methionine production—A critical review. Applied Microbiology and Biotechnology. 2014;**98**(24):9893-9914

[24] Goldflus F, Ceccantini M, Santos W. Amino acid content of soybean samples collected in different Brazilian states: Harvest 2003/2004. Brazilian Journal of Poultry Science. 2006;**8**(2):105-111

[25] Kita Y, Nakamoto Y, Takahashi M, Kitamura K, Wakasa K, Ishimoto M. Manipulation of amino acid composition in soybean seeds by the combination of deregulated tryptophan biosynthesis and storage protein deficiency. Plant Cell Reports. 2010;**29**(1):87-95

[26] Rackis JJ, Anderson RL, Sasame HA, Smith AK, VanEtten CH. Soybean amino acids, amino acids in soybean hulls and oil meal fractions. Journal of Agricultural and Food Chemistry. 1961;**9**(5):409-412

[27] Mato JM, Martinez-Chantar ML, Lu SC. Methionine metabolism. In: Amino Acids in Human Nutrition and Health. Wallingford, United Kingdom: CAB International; 2012. pp. 173-188

[28] Kredich NM. Regulation of cysteine biosynthesis in *Escherichia coli* and *Salmonella typhimurium*. In: Amino Acids: Biosynthesis and Genetic Regulation. Reading, Massachusetts: Addison-Wesley Publishing Company, Inc; 1983. pp. 115-132

[29] Fuller MF. Amino acid requirements for maintenance, body protein accretion and reproduction in pigs. In: Amino Acids in Farm Animal Nutrition. Wallingford, United Kingdom: CAB International; 1994. pp. 155-184

[30] Brosnan JT, Brosnan ME. The sulfur-containing amino acids: An overview. The Journal of Nutrition. 2006;**136**(6):1636S-1640S

[31] Jankowski J, Kubińska M, Zduńczyk Z. Nutritional and immunomodulatory function of methionine in poultry diets—A review. Annals of Animal Science. 2014;**14**(1):17-32

[32] D'Mello JPF. Responses of growing poultry to amino acids. In: Amino Acids in Farm Animal Nutrition. Wallingford, United Kingdom: CAB International; 1994. pp. 205-244

[33] Herrmann K. The common aromatic biosynthetic pathway. In: Amino Acids: Biosynthesis and Genetic Regulation. Reading, Massachusetts: Addison-Wesley Publishing Company, Inc; 1983

[34] Allee GL, Kerley MS. Modifications in soybean seed composition to enhance animal feed use and value: Moving from a dietary ingredient to a functional dietary component.

Journal of Agrobiotechnology Management & Economics. 2003. Vol 6. Number 1 & 2. Article 5. Available from: http://agbioforum.org/v6n12/v6n12a05-kerley.htm

[35] Wu G. Dietary requirements of synthesizable amino acids by animals: A paradigm shift in protein nutrition. Journal of Animal Science and Biotechnology. 2014;**5**(1):34

[36] Ingenbleek Y, Kimura H. Nutritional essentiality of sulfur in health and disease. Nutrition Reviews. 2013;**71**(7):413-432

[37] Levine RL, Mosoni L, Berlett BS, Stadtman ER. Methionine residues as endogenous antioxidants in proteins. Proceedings of the National Academy of Sciences of the United States of America. 1996;**93**(26):15036-15040

[38] Moskovitz J. Methionine sulfoxide reductases: Ubiquitous enzymes involved in antioxidant defense, protein regulation, and prevention of aging-associated diseases. Biochimica et Biophysica Acta. 2005;**1703**(2):213-219

[39] Catoni GL. S-Adenosylmethionine—A new intermediate formed enzymatically from L-methionine and adenosinetriphosphate. The Journal of Biological Chemistry. 1953;**204**(1): 403-416

[40] Fontecave M, Atta M, Mulliez E. S-adenosylmethionine: Nothing goes to waste. Trends in Biochemical Sciences. 2004;**29**(5):243-249

[41] Bunchasak C. Role of dietary methionine in poultry production. The Journal of Poultry Science. 2009;**46**(3):169-179

[42] Conde-Aguilera JA, Cobo-Ortega C, Tesseraud S, Lessire M, Mercier Y, van Milgen J. Changes in body composition in broilers by a sulfur amino acid deficiency during growth. Poultry Science. 2013;**92**(5):1266-1275

[43] Wu B, Cui H, Peng X, Fang J, Cui W, Liu X. Effect of methionine deficiency on the thymus and the subsets and proliferation of peripheral blood T-cell, and serum IL-2 contents in broilers. Journal of Integrative Agriculture. 2012;**11**(6):1009-1019

[44] Jessop CE, Chakravarthi S, Watkins RH, Bulleid NJ. Oxidative protein folding in the mammalian endoplasmic reticulum. Biochemical Society Transactions. 2004;**32**(Pt 5): 655-658

[45] Wylie LM, Robertson GW, Macleod MG, Hocking PM. Effects of ambient temperature and restricted feeding on the growth of feathers in growing turkeys. British Poultry Science. 2001;**42**(4):449-455

[46] Bonato MA, Sakomura NK, Siqueira JC, Fernandes JBK, Gous RM. Maintenance requirements for methionine and cysteine, and threonine for poultry. South African Journal of Animal Science. 2011;**41**(3):209-222-222

[47] Lewis AJ. Methionine-cysteine relationships in pig nutrition. In: Absorption and Utilization of Amino Acids. Wallingford, United Kingdom: CAB International; 2003. pp. 143-156

[48] Liu K. Soybeans: Chemistry, Technology, and Utilization. New York, NY: Chapman & Hall; 1997

[49] Sleper DA, Poehlman JM. Breeding Field Crops. 5th ed. Ames, Iowa: Blackwell Publishing Professional; 2006

[50] Hermann FJ. A Revision of the Genus Glycine and its Immediate Allies. Washington, D.C: U.S. Department of Agriculture; 1962. 84 p

[51] Howell RW, Brim CA, Rinne RW. The plant geneticist's contribution toward changing lipid and amino acid composition of soybeans. Journal of the American Oil Chemists' Society. 1972;**49**(1):30-32

[52] Mahmoud AA, Natarajan SS, Bennett JO, Mawhinney TP, Wiebold WJ, Krishnan HB. Effect of six decades of selective breeding on soybean protein composition and quality: A biochemical and molecular analysis. Journal of Agricultural and Food Chemistry. 2006;**54**(11):3916-3922

[53] Panthee D, Pantalone VR. Registration of soybean germplasm lines TN03-350 and TN04-5321 with improved protein concentration and quality. Crop Science. 2006;**46**:2328-2329

[54] Takahashi M, Uematsu Y, Kashiwaba K, Yagasaki K, Hajika M, Matsunaga R, et al. Accumulation of high levels of free amino acids in soybean seeds through integration of mutations conferring seed protein deficiency. Planta. 2003;**217**(4):577-586

[55] Meinke DW, Chen J, Beachy RN. Expression of storage-protein genes during soybean seed development. Planta. 1981;**153**:130-139

[56] Wolf WJ. Soybean protein nomenclature: A progress report. Cereal Science Today. Number 3. 1969;**14**:75-129. Available from: https://pubag.nal.usda.gov/catalog/31396

[57] Wilcox JR, Shibles RM. Interrelationships among seed quality attributes in soybean. Crop Science. 2001;**41**(1):11-14

[58] Paek NC, Imsande J, Shoemaker RC, Shibles R. Nutritional control of soybean seed storage protein. Crop Science. 1997;**37**(2):498-503

[59] Thakur M, Hurburgh CR. Quality of US soybean meal compared to the quality of soybean meal from other origins. Journal of the American Oil Chemists' Society. 2007;**84**(9):835-843

[60] Panthee DR, Pantalone VR, Saxton AM, West DR, Sams CE. Genomic regions associated with amino acid composition in soybean. Molecular Breeding. 2006;**17**:79-89

[61] Fallen B, Hatcher C, Allen F, Kopsell D, Saxton A, Chen P, et al. Soybean seed amino acid content QTL detected using the universal soy linkage panel 1.0 with 1,536 SNPs. Journal of Plant Genome Sciences. 2013;**1**(3):68-79. Agronomy and Horticulture—Faculty Publications. Available from: https://digitalcommons.unl.edu/agronomyfacpub/807

[62] Warrington CV, Abdel-Haleem H, Hyten DL, Cregan PB, Orf JH, Killam AS, et al. QTL for seed protein and amino acids in the Benning × Danbaekkong soybean population. Theoretical and Applied Genetics. 2015;**128**(5):839-850

[63] Li X, Tian R, Kamala S, Du H, Li W, Kong Y, et al. Identification and verification of pleiotropic QTL controlling multiple amino acid contents in soybean seed. Euphytica. 2018;**214**(6):93

[64] Panthee DR, Pantalone VR, Sams CE, Saxton AM, West DR, Orf JH, et al. Quantitative trait loci controlling sulfur containing amino acids, methionine and cysteine, in soybean seeds. Theoretical and Applied Genetics. 2006;**112**(3):546-553

[65] Townsend JA, Thomas LA, Kulisek ES, Daywalt MJ, Winter KRK, Altenbach AB. Improving the quality of seed proteins in soybean. In: Proceedings of the 4th Biennial Conference on Molecular and Cellular Biology of Soybean. Ames, Iowa: Iowa State University; 1992. p. 4

[66] Kerr P. Utilization and Quality of Identity Preserved Oilseed Co-Products. Presentation, Institute of Food Technologists; Symposium. New Orleans, LA; 1996

[67] Kim W-S, Krishnan HB. Impact of co-expression of maize 11 and 18 kDa δ-zeins and 27 kDa γ-zein in transgenic soybeans on protein body structure and sulfur amino acid content. Plant Science. 2019;**280**:340-347

[68] Falco SC, Guida T, Locke M, Mauvais J, Sanders C, Ward RT, et al. Transgenic canola and soybean seeds with increased lysine. Bio/Technology. 1995;**13**(6):577

[69] El-Shemy H, Khalafalla M, Fujita K, Ishimoto M. Improvement of protein quality in transgenic soybean plants. Biologia Plantarum. 2007;**51**:277-284

[70] Schmidt MA, Barbazuk WB, Sandford M, May G, Song Z, Zhou W, et al. Silencing of soybean seed storage proteins results in a rebalanced protein composition preserving seed protein content without major collateral changes in the metabolome and transcriptome. Plant Physiology. 2011;**156**(1):330-345

[71] Haghiri F. Influence of macronutrient elements on the amino aicd composition of Soybean plants. Agronomy Journal. 1966;**58**:609-612

[72] Kapoora AC, Gupta YP. Changes in proteins and amino acids in developing soybean seed and effect of phosphorus nutrition. Journal of the Science of Food and Agriculture. 1977;**28**(2):113-120

[73] Arora SK, Luthra YP. Relationship between sulphur content of leaf with methionine, cysteine and cysteine contents in the seeds of *Phaseolus aureus* L. As affected by S, P, and N application. Plant and Soil. 1971;**34**(1):91-96

[74] Gayler KR, Sykes GE. Effects of nutritional stress on the storage proteins of soybeans. Plant Physiology. 1985;**78**(3):582-585

[75] Gioseffi E, de Neergaard A, Schjoerring JK. Interactions between uptake of amino acids and inorganic nitrogen in wheat plants. Biogeosciences. 2012;**9**(4):1509-1518

[76] Teixeira WF, Fagan EB, Soares LH, Umburanas RC, Reichardt K, Neto DD. Foliar and seed application of amino acids affects the antioxidant metabolism of the Soybean crop. Frontiers in Plant Science. 2017;**8**. Available from: https://www.frontiersin.org/articles/10.3389/fpls.2017.00327/full

[77] Carrera CS, Reynoso CM, Funes GJ, Martínez MJ, Dardanelli J, Resnik SL. Amino acid composition of soybean seeds as affected by climatic variables. Brazilian Agricultural Research. 2011;**46**(12):1579-1587

[78] Wolf RB, Cavins JF, Kleiman R, Black LT. Effect of temperature on soybean seed constituents: Oil, protein, moisture, fatty acids, amino acids and sugars. Journal of the American Oil Chemists' Society. 1982;**59**(5):230-232

[79] Cai T, Chang K-C. Processing effect on soybean storage proteins and their relationship with tofu quality. Journal of Agricultural and Food Chemistry. 1999;**47**(2):720-727

[80] Mujoo R, Trinh DT, Ng PKW. Characterization of storage proteins in different soybean varieties and their relationship to tofu yield and texture. Food Chemistry. 2003;**82**(2):265-273

[81] Weng TM, Chen MT. Changes of protein in Natto (a fermented soybean food) affected by fermenting time. Food Science and Technology Research. 2010;**16**(6):537-542

[82] Lusas EW. Soybean processing and utilization. In: Soybeans: Improvement, Production, and Uses. 3rd ed. Madison, WI: ASA, CSSA, SSSA; 2004. pp. 949-1036

[83] Johnson L, Smith K. Soybean Processing. United Soybean Board. Soybean Meal INFO Center. Fact Sheet; 2019. 7 pages. Available from: https://www.soymeal.org/wp-content/uploads/2018/04/soybean_processing.pdf

[84] Parsons CM, Hashimoto K, Wedekind KJ, Han Y, Baker DH. Effect of overprocessing on availability of amino acids and energy in soybean meal. Poultry Science. 1992;**71**(1):133-140

[85] Taira H. Studies on amino acid contents of processed soybean: Part X. the influence of added sugars on the heat destruction of the basic and sulphur containing amino acids in soybean products. Agricultural and Biological Chemistry. 1966;**30**(9):847-855

[86] Araba M, Dale NM. Evaluation of protein solubility as an indicator of overprocessing soybean meal. Poultry Science. 1990;**69**(1):76-83

[87] Batal AB, Douglas MW, Engram AE, Parsons CM. Protein dispersibility index as an indicator of adequately processed soybean meal. Poultry Science. 2000;**79**(11):1592-1596

[88] Dilger RN, Sands JS, Ragland D, Adeola O. Digestibility of nitrogen and amino acids in soybean meal with added soyhulls. Journal of Animal Science. 2004;**82**(3):715-724

[89] Husek P, Simek P. Advances in amino acid analysis. Analytical and Bioanalytical Chemistry. 2001;**19**(9):986-999

[90] Tristram GR, Rattenbury JM. The development of amino acid analysis. In: Amino Acid Analysis. Chichester, UK: Ellis Horwood; 1981. pp. 16-36

[91] Williams AP. Recent developments in amino acid analysis. In: Amino Acids in Farm Animal Nutrition. Wallingford, United Kingdom: CAB International; 1994. pp. 11-36

[92] Jajić I, Krstović S, Glamočić D, Jakšić SM, Abramović BF. Validation of an HPLC method for the determination of amino acids in feed. Journal of the Serbian Chemical Society. 2013;**78**(6):839-850

[93] Oomah BD, Voldeng H, Fregeau-Reid JA. Characterization of soybean proteins by HPLC. Plant Foods for Human Nutrition. 1994;**45**:251-263

[94] Malmer MF, Schroeder LA. Amino acid analysis by high-performance liquid chromatography with methanesulfonic acid hydrolysis and 9-fluorenylmethylchloroformate derivatization. Journal of Chromatography. A. 1990;**514**:227-239

[95] Baianu I, Prisecaru V. Novel NIR spectroscopy correlation approach to amino acid analysis of soybean proteins for composition improvements. Nature Precedings. University of Illinois at Urbana. 2011. Available from: http://dx.doi.org/10.1038/npre.2011.6231.1

[96] Kovalenko IV, Rippke GR, Hurburgh CR. Determination of amino acid composition of soybeans (Glycine max) by near-infrared spectroscopy. Journal of Agricultural and Food Chemistry. 2006;**54**(10):3485-3491

[97] Pazdernik DL, Killam AS, Orf JH. Analysis of amino and fatty acid composition in soybean seed, using near infrared refelctance spectroscopy. Agronomy Journal;**89**:679-685

Chapter 3

Improving Seed Quality of Soybean Suitable for Growing in Europe

Aleksandra Sudarić, Maja Matoša Kočar,
Tomislav Duvnjak, Zvonimir Zdunić and
Antonela Markulj Kulundžić

Additional information is available at the end of the chapter

http://dx.doi.org/10.5772/intechopen.89922

Abstract

The potential of soybean for food, feed, and pharmaceutical industry arises from the composition of its seed. Since European countries import 95% of the annual demand for soybean grains, meal, and oil, causing an enormous trade deficit, the governments in Europe had started to introduce additional incentives to stimulate soybean cropping. To rebalance the sources of soybean supply in the future, production must be followed by continuous research to create varieties that would make European soybean more appealing to the processing industry and profitable enough to satisfy European farmers. This chapter is giving an overview of the European soybean seed quality research and an insight into soybean seed quality progress made at the Agricultural Institute Osijek, Croatia. The studies presented are mainly considering maturity groups suitable for growing in almost all European regions. The most important traits of soybean seed quality discussed are protein content and amino acid composition, oil content and fatty acid composition, soluble sugars, and isoflavones. Defining quality traits facilitates the parental selection in breeding programs aiming to improve the added value properties of final soybean products and enables the exchange of materials between different breeding and research institutions to introduce diversity, which is a prerequisite for genetic advance.

Keywords: soybean, seed quality, breeding, variability, Europe, chemical composition

1. Introduction

Soybean (*Glycine max* (L.) Merr.) is the main oilseed crop of the world [1], a staple crop for protein-rich food and feed as well as a significant source of nutraceutical compounds with many different medical benefits [2]. One of the major health-promoting traits of soybean seed is the proposed ability to reduce the risk of metabolic disorders, cardiovascular diseases and cancers [3, 4]. Even though the majority of soybean in the European Union (EU) is used as poultry and pork feed, and to a less extent for feeding dairy cows [5], the high nutritional value of the seed and its health-promoting traits are making soybean more and more appealing for human consumption. In the last decades (1998–2017), most of the world's soybean seed was produced in the Americas (85.6%), whereas Europe produced only 1.6% [1]. Moreover, European countries import 95% of the annual demand for soybean grains, meal and oil from overseas, causing an enormous trade deficit, so the governments in Europe had started to introduce additional incentives aiming to stimulate soybean cropping [6, 7]. As a result, the harvested area in Europe has been continuously increasing over the past few years [1]. Furthermore, EU has a relatively high demand for non-genetically modified (non-GM) soybean in comparison to other parts of the world, and food/feed ingredients containing more than 0.9% of GM material must be labelled. This generated a system of segregation and identity preservation (IP) that ensures the identity of non-GM soybean is preserved through the entire supply chain. If we consider that EU imports about 2.7 million tons of non-GM IP soybean meal equivalent yearly and that the premium for non-GM IP is 20–30% of the price of non-segregated soybean while more than 80% of soybean area worldwide is planted with GM soybean [5], inciting the production of non-GM soybean in Europe becomes a matter of high importance. In order to keep the positive trends and rebalance the sources of soybean supply in the future, growing soybean production must be followed by continuous and intensive research in order to create improved varieties which would make European soybean more appealing to the processing industry as well as profitable enough to satisfy European farmers. Furthermore, because the frequency of adverse weather events has been increasing over the last 20 years [8], continuous and intensive research is necessary to create stable European varieties with high seed quality which would become an integral part of conservation agriculture.

Regardless of the final goal, the prerequisite of successful crop improvement is genetic variability of the traits of interest. Assessment of genetic diversity is necessary for germplasm characterisation, conservation, utilisation and the establishment of breeding programmes [9]. If sources of genetic variability are available, the genetic advance can be achieved not only with genetic engineering but also with continuous breeding efforts using conventional hybridization and selection methods together with modern chemical, biochemical and genetic analyses [10]. Conventional soybean improvement starts with the creation of a large recombinant inbred line population of significant diversity by hybridization of chosen parental components carrying the traits of interest. According to Burton [11], genetically distant elite parental lines have the greatest chance of producing superior progeny. Genetic characterisation and evaluation of divergence of parental material prior to hybridization can be made based on the pedigree [12], by measuring and analysing the variability of qualitative and quantitative morphological properties [13], on the basis of biochemical properties [14, 15] and with the help of molecular markers [16, 17]. All phenotype determination methods must consider the interaction of genotype and environment, whereas the efficacy and reliability of molecular markers are based on the ability of direct and rapid determination of genotype divergence excluding the environmental influences. Even if molecular methods of determining the differences in genetic constitutions are not available, creating a diverse population is enabled by the positive correlation between phenotypic variability and genetic divergence [18]. Therefore, phenotype evaluation is crucial for successful crop improvement.

The process that negatively affects crop improvement is the loss of genetic diversity as a consequence of human activity and the influence of the environment. This process, which Harlan [19] called genetic erosion, appears because of the replacement of diverse indigenous populations with modern, new, uniform cultivars and hybrids, and it causes a considerable threat to the production of food and hence the survival of humans. A narrow genetic base has been identified in most soybean germplasm studies [20, 21]. However, Hahn and Würschum [22] noted that the genetic base of middle European genotypes was not as narrow as expected because of the unsystematic phenotype selection. If narrowing of the genetic base is not stopped, it can result in complete crop destruction due to the lack of tolerance to adverse abiotic or biotic factors. This is why it is important to take all the measures necessary for preserving genetic diversity not only as it is crucial for crop improvement [23] but also to preserve biodiversity which ensures natural sustainability for all life forms. Furthermore, it is crucial to ensure the availability of this biodiversity to breeders, researchers and producers in order to ensure the genetic advance of cultivars in the future. As in other plant species, genetic diversity for soybean is preserved by creating germplasm collections on a local level and gene banks on a global level. Creating large and diverse germplasm collections as genetic resources necessary for the production of high-yielding and high-quality, commercially important cultivars is only possible if there is a continuous research. Since the commercial significance of soybean emanates mainly from the chemical composition of its seed, further improving the genetic basis of soybean seed quality should nowadays be as important as increasing the seed yield. Higher seed yield means more food produced on the same land area, but when 2 billion people worldwide are known to suffer from malnutrition [24], the quality of that food is of concern as well. This is why crops with high nutritional value, such as soybean, can greatly contribute to improving human and animal diets and health, as well as provide quality stock for pharmaceutical and functional food industries.

This chapter is giving insight into soybean seed quality research beneficial for defining the quality traits in available germplasm and choosing parental components in the soybean breeding programmes aiming to improve the added value properties of final soybean products. The studies presented here are mainly focused on maturity groups (MGs) 00 to II commonly sown in Central and South-eastern Europe but suitable for growing in almost all European regions [7]. The most important traits of soybean seed quality discussed here are protein content and amino acid composition; oil content and fatty acid composition; content and composition of soluble sugars, especially oligosaccharides; and content and composition of isoflavones.

2. Protein content and amino acid composition

Soybean protein, taking up 40% of the dry seed weight (DW) on average [25], is highly valued for food and feed because of its amino acid composition and a high digestibility [26]. Although it is thought to be the equivalent of animal protein [27], soybean proteins are deficient in sulphur-containing amino acids of which methionine and cysteine are considered the most limiting in animal feed [28]. Nevertheless, after appropriate heat treatment to reduce protease inhibitor activity, soybean proteins are considered to be superior to the proteins of other legumes in their growth-promoting properties [29]. Soybean seed proteins are made up from four major fractions: 2S, 7S, 11S and 15S, with 7S (glycinin) and 11S (ß-conglycinin) being the most abundant [30]. The ratio between 7S and 11S subunits has an important role in determining the functional properties of food made from soybean [31] and protein quality in soybean as glycinin has more sulphur-containing amino acids (cysteine and methionine) than ß-conglycinin [32]. Furthermore, glycinin is considered to lower the cholesterol levels in human serum [33], and it is important for tofu gel formation [34], whereas α subunit of ß-conglycinin is identified as one of the major allergenic proteins in soybean [32]. The nutritional value, utilization and digestibility of soybean proteins are further affected by bioactive compounds with toxic and/or anti-nutritional properties such as lipoxygenases, lectins, urease, the Kunitz trypsin inhibitor and the Bowman-Birk trypsin inhibitor [35], causing digestive and metabolic diseases in animals [36]. Some of these anti-nutritional factors can be destroyed or inactivated by heat treatment and some by supplemental enzymes, but others are unaffected by the methods applied commercially [37]. Nevertheless, in 2018, soybean represented 70% of protein meal used worldwide [38]. Defatted soybean meal has the highest level of crude protein among plant-based protein sources [26], and it is the main source of protein in commercial feed mixtures for poultry, livestock and fish farms [39]. For the production of these nutritionally valuable proteins, soybean needs less land area and less N than wheat and other small grains, maize, rice, fruits and vegetables. The emergence of soybean as a dominant global crop resulted in a partial offsetting of the dramatic increases in global soil nutrient withdrawals occurring as a result of the expansion of agriculture over newly cultivated land and rising yields in the recent decades while, at the same time, slightly increasing the overall protein content of the global harvest [40]. In the last several decades, the dominance of the soybean on the world crop market was further enhanced by the increase in meat consumption and livestock production, which caused the increase of the demand for high-protein materials for livestock feed in Europe as well [41]. It was estimated that global meat and milk demand would have an increase of 57 and 48% respectively, between 2005 and 2050, due to the fast-growing population and rising incomes in developing countries [42], while livestock production was estimated to increase by 21% between 2010 and 2025 [43]. Plant protein production in the EU, on the other hand, does not follow this increase, so almost 70% of domestic needs are covered by imported plant protein [44]. Because importing 1 kg of dried soybean meal to the European Union from South America is associated with 11.65 kg CO_2 equivalent emissions [45] and because the price of proteins on the world market is continuously rising [46], this imposes major environmental, economic and social problems. Therefore, increasing local plant protein production and superseding the import of feed protein would not only increase

Europe's protein self-sufficiency but also decrease the negative environmental footprint of animal production [47]. Furthermore, including legumes such as soybean in crop rotation dominated mainly by cereals and non-legume oilseeds, which is the case in Europe [44], can have positive effects on soil quality and agrobiodiversity and can contribute to reducing nitrous oxide emissions and nitrate-N leaching [48]. These environmental benefits, together with benefits arising from the better balance of EU agriculture and trade, are the main reasons the provisions of the new Common Agriculture Policy (CAP) include the promotion of protein crops in Europe as a priority [49].

In general, increasing plant protein production can be done by increasing the area under high-protein crops or by increasing the protein content by breeding. Even some management practices can influence protein content. For example, minimum tillage and herbicide application can decrease the protein content in comparison to conventional tillage and no herbicides [50], and irrigation at the beginning of pod formation and during seed-filling can result in protein content increase [51–53]. Furthermore, Foroud et al. [54] and Bouniols et al. [55] noted that protein content was highest when irrigation was applied after the flowering, whereas continuous water supply decreased it. According to Afza et al. [56], nitrogen (N) fertilization during the seed-filling stage increased the protein content, and earlier N application, e.g. in flowering, did not affect the protein content [57], whereas, sulphur-containing amino acid composition fluctuated depending on the nitrogen source [58] and on the availability of reduced forms of sulphur as well [59]. Zimmer et al. [60] reported that protein content of non-inoculated soybeans was significantly lower than protein content of the inoculated soybeans, and Vollmann et al. [61] reported that N fertilization at the flowering stage was superior to both the control and rhizobium inoculation in increasing seed protein content, but as symbiotic N fixation is a highly complex phenomenon, the difference in protein content could have been the result of many different factors and their interactions which demand further investigation. Although correct management practices can increase the protein yield of soybean crops, improving the genetic base of protein content is the right approach for providing high-quality soybean cultivars. Furthermore, because the agroclimatic conditions in Europe are not ideal for widespread cultivation of protein crops like soybeans [62], increasing protein yield per area of arable land by plant breeding is the only efficient way of decreasing the plant protein deficit in Europe in which genetic diversity and population structure have a key role [22].

The soybean protein content is a quantitative trait inherited polygenetically, and it ranges between 30 and 50% on a dry weight basis [10]. Besides earlier mentioned management practices (M), protein content is influenced by genotype (G), environment (E) and interactions G × E × M, but the contribution of each is still not well established [50, 63–67]. The G and E interaction is one of the main problems in genotype selection as well as in the recommendation of cultivars. Because of this, estimating the broad sense trait heritability is essential for successful genetic advance and the larger the estimated value of this parameter is, the greater will be the chance of success with selection [68]. Broad-sense heritability assessments for soybean vary from middle, which suggests variability due to the combination of genetic and environmental factors, to high, where genetic factors are more pronounced in determining the protein phenotypes [67, 69, 70]. The influence of the environment on protein content has been researched by many authors. According to Popović et al. [71] and Josipović et al.

[66], protein content is higher in years with lower average temperatures and more precipitation during pod formation and seed-filling stages. On the other hand, Hurburgh [72] and Nian et al. [73] reported reduced protein content in northern regions of soybean cultivation with lower temperatures and higher amounts of precipitation, which can be the result of reduced symbiotic N fixation and synthesis of proteins due to the low root-zone temperatures [74]. Gibson and Mullen [75] reported that high daily temperatures in combination with high night temperatures during the period from seed-filling to maturity can cause positive linear relationship of average temperatures and protein content. Furthermore, Dornbos and Mullen [76] reported that high temperatures in combination with drought in seed-filling period can result in significant protein content rise, whereas Vollmann et al. [61] and Matoša Kočar et al. [77] noticed that seed protein content was highest for soybean crops grown under moderately dry conditions and high temperature during the seed-filling stage. Such contradictory reports are a major concern in plant breeding since they complicate decision-making process during selection. Breeding for increased protein content is further complicated by negative correlation between protein content and yield reported by many authors [70, 78, 79], although there are studies indicating no significant correlation between those two parameters [80, 81]. Furthermore, Vollmann et al. [61] suggested that even moderately negative correlation between seed protein content and seed yield, means the selection of breeding lines with both improved protein content and acceptable yield level should be possible. On the other hand, the relationship between protein content and oil content is almost always significantly negative [61, 70, 81] which is caused by either pleiotropic effect or linkage [82] where every 2% protein content increase usually decreases oil content by 1% [83]. Apart from protein content, the content of sulphur-containing amino acids should be considered as it affects the nutritional value of soybean meal, so increasing the content of both would be beneficial. However, Wilcox and Shibles [84] determined a weak and inverse correlation between protein content and methionine and cysteine levels which means protein quality decreases with the increase in protein content, so Paek et al. [58] concluded that it would be beneficial to focus on improving protein quality over protein content. Nevertheless, Burton et al. [85] noticed no significant changes in methionine content among cycles of recurrent selection for high protein. Furthermore, Krober and Cartter [86] reported a positive relationship between methionine content and protein content suggesting the possibility of combining high-yield, high-quality protein soybean lines.

As mentioned earlier, improving any trait of interest should start with screening the available materials. In Europe, soybean seed content was researched by many authors over the years. Vollmann et al. [61] determined a considerable variation in protein content of early maturing genotypes studied during 6 years in Austria. In their study, genotype × treatment and genotype × year interactions were of a much lower magnitude than the effect of genotype on protein content [61]. Sudarić et al. [63] researched 14 MG 0-I soybean genotypes at three locations in eastern Croatia during 5 years and reported highly significant effects of genotype, year, location, genotype × year, genotype × location interactions as well as a highly significant effect of the interaction between genotype, year and location with year effects being larger than location effects. Protein content range in the research by Popović et al. [71] was 36.65–37.66%, and average value was 37.15%. Fogelberg and Recknagel [87] determined the protein content range of early matured, well-adapted soybean cultivars evaluated at three sites in Germany during 3 years to be 37–43%, while Pannecoucque et al. [81] reported a protein content range of 35.5–43.3% in 14 early maturing soybean varieties tested in two consecutive years in two locations in Belgium. In a study by Kurasch et al. [70] two types of European soybean varieties that maximise protein yield were distinguished: one that produces high grain yields per hectare with protein contents around 39–42% and the other with slightly lower protein yields than the first type but characterised by very high-protein contents (above 42%). Soybean breeding at the Agricultural Institute Osijek (Croatia) resulted in an increase of protein content over time, and this progress can be followed through three studies. The first one [88], conducted in the 1990s, researched 22 soybean cultivars (11 standard cultivars and 11 promising lines) whose protein content range was determined to be 35.9–38.4%. Afterwards, in a 3-year study (2001–2003), Vratarić et al. [89] tested 132 elite breeding lines and standard cultivars grouped according to their MG (00, 0, I) and reported that for all MGs, elite breeding lines had significantly higher values for protein content than their respective standards. Ranges determined were 35.2–38.7%, 36.2–39.6% and 38.7–41.2% for MG 00, 0 and I, respectively [89]. In the third research, Matoša Kočar et al. [77] determined significant variation for soybean seed protein content, influenced by genotype and year in a 3-year (2010–2012) screening study with 21 soybean genotypes (MG 00 to I) created at the Agricultural Institute Osijek (Croatia). The lack of significant difference in average protein content between elite breeding lines created at the Agricultural Institute Osijek and standard cultivars (**Table 1**) can be explained with the significant progress already achieved in previous selection cycles [89]. Nevertheless, wide ranges of variation for protein content in genotypes from the Agricultural Institute Osijek indicate there is still a room for improvement [77]. Although, according to Sato et al. [90], most of the world is focused towards increasing soybean seed oil content and oil yield, average protein content is at the world level higher than protein content reported for Central European soybean varieties [61]. For example, Bueno et al. [64] reported that average protein content value in 18 Brazilian soybean genotypes was 42.44% (40.20–44.49%), Sharma et al. [91] reported average protein content values being 41.4% (39.40–44.40%) in eight genotypes in India, while Ramteke et al. [92] reported somewhat lower average protein content (40.23%) determined in 92 soybean varieties in India with values ranging from 37.69 to 42.74%. Furthermore, protein content determined

MG	Elite breeding lines		Standard cultivars		t	P
	Range	Mean	Range	Mean		
	Protein (% DW)					
00	35.86–43.82	39.87	35.52–43.55	39.44	0.37^{ns}	$P > 0.05$
0	35.83–44.33	40.42	34.14–42.48	39.08	1.28^{ns}	$P > 0.05$
I	36.02–44.04	40.19	36.34–42.70	39.07	1.40^{ns}	$P > 0.05$

The means were tested with Student's t-test with ns indicating no significant difference.

Table 1. The 3-year (2010–2012) protein content range and mean for different maturity group (MG) elite breeding lines created at the agricultural institute Osijek in comparison to standard cultivars.

in the whole US Department of Agriculture (USDA) soybean germplasm collection ranged from 34.1 to 56.8% [93]. Higher average protein contents in non-European studies are maybe due to the fact that soybean breeding is still at a relatively low level in Europe compared to the USA or Asia because most of the soybean demand in Europe has been relying on import [44], so developing high-quality genotypes was not of great economic importance. Therefore, European breeding programmes would benefit from introducing foreign accessions with high-protein contents to be used as parental components in crosses, providing these accessions are not transgenic, since that is not socially acceptable in Europe [70].

If sources of variation for high-protein content are available, the next step is choosing the most efficient breeding method. Sato et al. [90] reported that crossing adapted elite genotypes with advanced high-protein donors exhibiting a seed protein content of approximately 470 g/kg protein improved protein content in progeny more than crossing adapted genotypes and food-grade cultivars. Increasing the protein content from 46.3% in the initial parental population to 48.4% was achieved after six cycles of selection, without significantly reducing yield [78]. Furthermore, Thorne and Fehr [94] concluded that three-way crosses were superior to two-way crosses for creating high-protein, high-yielding lines. Wehrmann et al. [95] reported that selection for high protein between two backcross generations increased protein content while maintaining seed yield. Wilcox and Cavins [96] were able to create high-protein BC_3 line with the yield at the same level as the recurrent parents, but it took them 20 years because it was necessary to select for high protein between each backcross. Cober and Voldeng [80] evaluated single-cross and backcross breeding methods of achieving high-yield, high-protein soybean genotypes and found no significant differences between them. Selecting appropriate genotypes during the breeding process can be made easier and more efficient with marker-assisted selection (MAS), especially since quantitative trait loci (QTL) for proteins have been mapped to locations on all chromosomes. QTL mapping and genome-wide association studies (GWAS) have identified 252 QTL associated with soybean protein and distributed on all 20 soybean chromosomes [97]. Brummer et al. [98] evaluated eight soybean populations from the Midwestern USA for genetic markers linked to seed protein content detecting a significant association between markers and traits and identifying environmentally stable and sensitive QTL. Sebolt et al. [99] successfully integrated *G. soja* alleles for high-protein content into *G. max* background by backcrossing. Genetic marker alleles linked to the QTL allele from *G. soja* on linkage group (LG) I were significantly associated not only with higher protein content but with lower oil content, reduced yield, smaller seeds, taller plants and earlier maturity. Hwang et al. [100] used GWAS to identify QTL controlling seed protein content in 298 soybean germplasm accessions exhibiting a wide range of protein content and identified 40 single-nucleotide polymorphisms (SNPs) in 17 different genomic regions significantly associated with seed protein. GWAS also resulted in narrowing the genomic region previously reported to contain protein content QTL [101] which is important because narrower GWAS-defined genome regions will allow more precise marker-assisted allele selection and will expedite positional cloning of the causal gene(s) [100]. Bandillo et al. [102] studied 12,000 accessions from the USDA soybean germplasm collection by using GWAS and had identified SNPs for protein and oil content with strong signals on chromosomes 20 and 15 with chromosome 20 region, previously reported to be important for protein and oil content, further narrowed so it contained only three plausible candidate genes. Besides genomic studies, identifying markers for protein content can be done using metabolomics, a non-targeted approach monitoring hundreds of metabolites, but there is still little knowledge about the association between metabolites and protein content in soybean seeds, especially in similar genetic backgrounds [103]. Nevertheless, Wang et al. [103] evaluated metabolic diversity in a soybean near-isogenic line (NIL) population derived from parents with contrasting seed oil contents comparing seed primary metabolites of high-protein/low oil lines, low-protein/high oil lines and their parents. Results indicated that metabolic profiles of all progeny lines could be discriminated based on protein and oil contents. All such molecular and genomic investigations help to reveal the genetic architecture of complex traits, protein content being one of them, and enable understanding of the genetic basis of trait variation which is crucial for improving seed quality as well as other important agronomic traits. Nonetheless, local cultivar testing is particularly important for the development of the crop, despite the wide adaptation of soybean, since there could be substantial site effects within regions [87] which mean it would be beneficial for the European soybean breeders to develop their own, local varieties with increased protein content, rather than relying on foreign introductions.

The next step after increasing soybean seed protein content could be breeding for more favourable amino acid composition and improved 11S/7S protein fraction ratio. For example, while the manipulation of other aspects of seed composition and processing may improve amino acid assimilation, increasing the relative proportion of methionine, lysine and threonine has become a goal in soybean breeding [104]. Almost 20 years ago, the economic benefit of improved essential amino acid content was estimated to be ~$5 per ton per 10% increase of any of the above-mentioned amino acids [105], and today the benefit could be even bigger. The knowledge of the molecular mechanism of amino acid biosynthesis in soybean was mainly limited to genetic mapping [106, 107], but recent studies have enhanced our understating of amino acid genetic architecture in soybean. For example, Vaughn et al. [108] conducted GWAS for methionine, threonine, cysteine and lysine and identified multiple loci associated with multiple amino acids across different populations not reported previously. Furthermore, Qiu et al. [109] predicted 12 candidate genes based on the synteny between 113 genes of sulphur-containing amino acid synthases and 33 related QTLs in soybean and bioinformatic analyses, and many QTL related to the 7S (β-conglycinin) and 11S (glycinin) fractions of soybean storage proteins have been identified on chromosomes 1, 3, 4, 6, 10, 13, 16, 17, 19 and 20 [110–112]. The 11S/7S ratio of soybean seed can vary greatly due to both genetic and environmental differences [113, 114], which means there is room for improvement. In a study by Žilić et al. [32], the ratio of 11S/7S proteins varied from 2.43 to 3.29 among the seven soybean varieties adapted to South-Eastern European conditions. Murphy and Resurreccion [113] reported the 11S/7S protein ratio to vary from 2.1 to 3.4 in 12 soybean varieties, while Mujoo et al. [114] determined this ratio to be 1.63–2.05 in 7 soybean varieties. Since 7S protein fraction is more favourable for food and feed, lowering 11S/7S ratio would be beneficial, but at the moment it is still far from a priority for European breeding programmes. Furthermore, according to Zhang et al. [115], breeding efforts for the improvement of soybean seed amino acids have rarely been reported, much less developed on a commercial scale, because methods used for quantifying amino acids in soybean, such as HPLC, are time-consuming and not cost-effective, therefore not suitable for high-throughput screening of a large number of lines [116].

3. Oil content and fatty acid composition

Soybeans represent 59% of the world's oilseed production and 29% of the total vegetable oil consumption in the world [117]. Oil from soybean seed can be found in margarine, salad dressings and cooking oils and in industrial products such as plastics and biodiesel fuel as well. The content of oil in soybean seed can range from 12 to 24% DW, while most commercial cultivars contain between 19 and 23% [110]. Soybean oil is composed of triacylglycerols (TAGs), which are products of fatty acid esterification [118]. Soybean oil fatty acids are responsible for nutritional value, stability and taste of soybean oil, and commodity soybean oil is in average made up from 13% palmitic acid (16:0), 4% stearic acid (18:0), 20% oleic acid (18:1), 55% linoleic acid (18:2) and 8% linolenic acid (18:3) [119]. Fatty acids in soybean can be saturated (palmitic and stearic acid) or unsaturated (oleic, linoleic and linolenic acid). Soybean oil intended for frying should, for example, contain 7% saturates, 60% oleic acid, 31% linoleic acid and 2% linolenic acid, whereas desired fatty acid composition of oils intended for industrial use is 11% saturates, 12% oleic acid, 55% linoleic acid and 22% linolenic acid. Nevertheless, the most desirable fatty acid phenotypes for soybean oil are those having the most applications for food, feed and industry, i.e. phenotypes with saturates reduced to less than 7%, linolenic acid reduced to less than 3% and oleic acid increased to more than 55% [93]. Lowering saturates makes oil more suitable for the food industry and consumers concerned about dietary health issues such as high cholesterol and increased risk of coronary heart disease usually associated with diets high in saturated fats [120]. However, as stearic acid is thought to be heart neutral [121], increasing its content could satisfy the market demand for healthier foods without compromising desired functional qualities contributed by saturated fats [122]. Unsaturated fatty acids can be monounsaturated (MUFA), such as oleic acid (n-9, omega-9), or polyunsaturated (PUFA), such as linoleic (n-6, omega-6) and linolenic (n-3, omega-3) acids. Oleic acid is not considered to be an essential fatty acid (EFA) since the human body can synthesise it in small amounts, but it is significant as a precursor of some EFAs [123]. On the other hand, polyunsaturated linoleic and linolenic acids are considered essential, supporting the cardiovascular, reproductive, immune and nervous systems, crucial for manufacturing and repairing cell membranes [123]. However, these PUFAs are susceptible to oxidation, so they reduce the shelf life of the oil, causing low stability at high cooking temperatures as well as off-flavours [124]. Oxidative stability of soybean oil generally decreases with the increasing degree of unsaturation [125] and can be assessed by the ratio between MUFA and PUFA contents (MUFA/PUFA) [126]. Soybean oil has poor oxidative stability compared to other vegetable oils, and its MUFA/PUFA is about 0.5 on average [126]. The average value for MUFA/PUFA in the research by Matoša Kočar et al. [127] was 0.4 suggesting these genotypes would give less stable final product, i.e. edible oil with shorter shelf life and lower stability for cooking at high temperatures [126]. Oxidative stability of soybean oil can be improved in two ways: by trans-isomer producing catalytic hydrogenation or by breeding for higher content of oleic acid (MUFA), which is also known to reduce cholesterol and reduce the risk of arteriosclerosis and heart disease [93, 124, 126]. On the other hand, diets rich in trans-fatty acids increase the risk of cardiovascular diseases [128], so increasing oil stability by hydrogenation is becoming more and more unpopular. Soybean oil with high-oleic acid (>70%) and low-linolenic acid content (<3%) is desirable for its oxidative stability and health benefits, with the linolenic acid content of <3% being the current market target to increase soybean oil functionality [129]. As oxidative stability of soybean oil can be assessed by MUFA/PUFA, nutritional quality can be assessed by the linoleic and linolenic acid ratio (LA/ALA) with an optimal value between 5 and 10:1 [126]. Three-year average LA/ALA reported in the study by Matoša Kočar et al. [127] was 7.36, which falls within the recommended range for vegetable oils suggested by Rani et al. [126], meaning that tested genotypes can produce nutritionally acceptable oil. On the other hand, very high LA/ALA is considered detrimental for human health, and so lowering it by breeding can help with the prevention against degenerative pathologies [130]. Even though the demand for vegetable oil is continuously increasing and this increase is expected to continue in the period from 2018 to 2027, as the production of oilseeds is mainly concentrated in few regions of the world, weather fluctuations and extreme whether having a negative effect on the yield impact oilseed market more than other major crop markets, causing market uncertainties and price volatility [131]. This, together with evident changes in consumer preferences concerning dietary value and functional properties of oil, is causing the world oilseed market to become more competitive, which encourages breeding programmes to focus not only on increasing oil yield but also on increasing oil quality to meet the demands of industry and end-users alike [132]. Furthermore, because European consumers still prefer GM-free oilseed end products, while the area planted with soybean, the main oilseed crop of the world [1], is more than 80% planted with GM soybean [5], European soybean breeders should focus on creating high-oil GM-free varieties with conventional breeding methods so that they are able to provide high-quality and healthy end products to European consumers. Seed oil content is a complex quantitative trait, controlled by multiple genes and affected by environmental factors [124, 133]. The broad-sense heritability of soybean oil content varied considerably in the range between 0.13 and 0.99 [7, 134–136] suggesting that trait inheritance is a function of G, E, M and G × E × M. For example, oil content can be in positive correlation with temperature [137, 138], but negative linear relationships [139] and quadratic relationships [140] were also reported. Water stress negatively impacts oil content [141], but irrigation at the beginning of pod formation and during seed filling has resulted in oil content decline [51–53]. Some other crop management factors (no-till, seed treatments, foliar N, fungicide and insecticide applications and rotation) had an overall positive effect on oil content because they were improving crop growing conditions through conserving or supplying water and nutrients or improving soil physical and chemical properties and protecting the crop from diseases [142]. On the other hand, a decline in oil content with increased N application [141] and lack of response [143] were reported as well. While monocropping negatively affected soybean oil composition in comparison to crop rotation with maize [144], early planting benefited oil [145–147]. Furthermore, Assefa et al. [142] reported that planting date had larger impacts at northern latitudes (40–45 N) than southern latitudes (30–35 N). Although the influence of M for oil content was found to be significant [144–147], in a comprehensive analysis of G, M and E factors influencing soybean yield quantity and quality across the US Corn Belt, Assefa et al.

[142] concluded that E is a dominant factor for the significant variability in seed composition and yield. According to Josipović et al. [66] and Popović et al. [71], the amount of oil in soybean seed was significantly higher in years with less precipitation and higher air temperatures at the time of pod formation and dry matter accumulation. This was also true for the average oil content determined by Matoša Kočar et al. [127], noting that hot and extremely humid conditions resulted in the lowest average amount of oil in soybean seed, whereas hot and dry conditions resulted in higher average amounts of oil. On the other hand, Vollmann et al. [61] reported that low precipitation rates and high temperatures did not favour oil synthesis and that, in genotypes of later maturity, oil synthesis was enhanced by late water availability. The significance of genotype and environmental factors for fatty acid content was determined by many authors as well [7, 61, 65, 67, 127, 146, 148]. Bellaloui et al. [146] noted that, among fatty acids, oleic acid was the most sensitive to environmental factors and palmitic and stearic acids were the least sensitive in clay soil, whereas stearic, linoleic and linolenic acids were the least sensitive in sandy loam soil. According to the same authors, cooler temperatures favour the synthesis of linolenic acid, whereas the synthesis of oleic acid is negatively affected as a result of the inverse relationship between them [146]. The same was true in the research done by Xue et al. [149], where increasing air temperature during pod fill significantly increased oleic acid and significantly decreased linoleic and linolenic acid contents which was later confirmed by Matoša Kočar et al. [129].

Together with understanding the sources of variation, knowing the strength and direction of relationships among different traits is of great value for breeding programmes as it enables the enhancement of more than one trait at the same time. Although significantly negative relationship between oil and protein content was reported in most researches [61, 70, 81, 84], in the research analysing variations from 21 studies conducted over 15 years (2002–2017), Assefa et al. [142] concluded that, when pooling data across all studies, there was no significant relationship between oil and protein content, but a tendency for a negative relationship was observed when plotting data separately for each of the studies evaluated in the database. Furthermore, oil content increased slowly with yield increase suggesting a positive relationship, but when relationships were investigated by study, 63% of studies supported a positive relationship, and the other 37% displayed a slightly negative relationship between seed yield and oil content [142]. Wilcox and Shibles [84] reported that oil increased by 1.9 g kg^{-1} for every 100 kg ha^{-1} increase in seed yield, whereas protein decreased by 15.6 g kg^{-1} for each 10 g kg^{-1} increase in oil, which is in accordance with determined significant positive correlation between oil content and yield and significant negative correlation between oil and protein contents. The analysis of the correlation between oil content and different fatty acid contents in eight early maturing soybean genotypes revealed a positive correlation between oil and saturated fatty acids but negative correlation between oil and unsaturated fatty acids [115]. Matoša Kočar et al. [127] reported a highly significant and positive correlation of oil with stearic acid and significant positive correlation with oleic acid but significant negative correlation with linoleic and linolenic acid. The results were somewhat different in the research conducted by Rani et al. [126] where oil was in highly significant positive correlation with stearic acid and linoleic acid but in negative correlation with oleic acid. The correlation between favourable oleic acid and linoleic acid was highly significant and negative in the research by Matoša Kočar et al. [127], the same as in some previous studies [126, 150], while the correlation between oleic acid and unfavourable linolenic acid was also highly significant and negative. These relationships indicate that it could be possible to create varieties with high oil amount, a high amount of oleic acid and low amount of linolenic acid, which is favourable for edible oil.

Whether the improvements in soybean oil yield and its quality are achieved by conventional breeding or genetic engineering [2, 119], having a favourable gene pool concerning the trait of interest is crucial. In Europe, the variability of oil content was studied by many authors [61, 70, 77, 88, 89, 127]. At the Agricultural Institute Osijek, breeding for soybean seed oil over time is presented by three studies. In the first one [88], the range for oil content determined in 22 soybean cultivars tested from 1993 to 1995 was 18.9–20.5%. The second one [89] tested 132 elite breeding lines and standard cultivars grouped according to their MG (00, 0, I) during 3 years (2001–2003). All three elite breeding line groups had significantly higher average oil contents than their respective standards. The range for MG 00 was 22.1–23.8%, for MG 0 22.5–23.4% and for MG I 21.9–22.7% [89]. The third research from the Agricultural Institute Osijek, conducted between 2010 and 2012 with eight MG 0 advanced soybean breeding lines, reported oil content ranges from 22.09 to 24.08% [77]. As it is evident from **Table 2**, there were no significant differences determined between the average oil content of elite breeding lines created at the Agricultural Institute Osijek in the last decade in comparison to standard cultivars. This is expected because significant progress has been achieved in earlier selection cycles

MG	Elite breeding lines		Standard cultivars		t	P
	Range	Mean	Range	Mean		
	Oil (% DW)					
00	20.59–25.48	23.29	21.01–23.81	22.54	1.49ns	P > 0.05
0	20.24–25.84	22.85	20.06–23.85	22.15	1.45ns	P > 0.05
I	20.93–26.40	23.24	20.66–24.40	22.40	1.76ns	P > 0.05
	Oleic acid (% oil)					
00	20.58–31.72	25.72	21.37–27.51	23.92	1.67ns	P > 0.05
0	19.51–28.91	23.59	18.74–24.84	21.38	2.68**	P < 0.01
I	20.27–28.05	24.78	19.48–26.83	23.54	1.30ns	P > 0.05
	Linolenic acid (% oil)					
00	4.41–7.04	5.99	5.36–8.27	6.93	2.23*	P < 0.05
0	4.99–8.65	7.15	6.97–7.99	7.59	2.54*	P < 0.05
I	5.10–7.98	6.59	5.37–8.73	7.17	1.37ns	P > 0.05

The means were tested with Student's t-test with * and ** indicating significant differences at $P \leq 0.05$ and $P \leq 0.01$, respectively, and ns indicating no significant difference.

Table 2. The 3-year (2010–2012) oil, oleic and linolenic acid content range and mean for different maturity group elite breeding lines created at the agricultural institute Osijek in comparison to standard cultivars.

and oil content in genotypes from Osijek is already at a relatively high level. Furthermore, as soybean seed is mostly used for animal feed in Croatia; further increasing oil content is not a priority. However, there were genotypes among the newer elite breeding lines that accumulated significantly higher amounts of oil than the rest, indicating their suitability for being used as parental components [77]. Somewhat lower oil content range values (18.55–18.73%) were reported by Kurasch et al. [70] researching 1008 $F_{5:8}$ recombinant inbred lines MG (MG 000–00) that showed good agronomic performance in Central Europe. Pannecoucque et al. [81] determined the oil content range being 20.3–24.1% in 14 very early (MG 0000–00) soybean genotypes selected from the European plant variety catalogue. Outside Europe, Bueno et al. [64] reported 22.1% average oil content determined in 18 Brazilian soybean genotypes with values ranging from 20.72 to 22.81%, and Sharma et al. [91] noted 16.3% average oil content for 8 genotypes in India ranging from 14 to 18.7%, whereas oil content determined in the whole US Department of Agriculture soybean germplasm collection ranged from 8.1 to 27.9% [93]. Variability of soybean seed fatty acid content in European commercial breeding programmes has seldom been researched, but in other parts of the world where soybean is considered the main crop feeding oil industry, many studies have been conducted. Significant variability of the fatty acid content was determined in eight advanced soybean breeding lines (MG 0) developed at the Agricultural Institute Osijek, Croatia [127]. The contents were 10.81, 5.78, 26.41, 49.75 and 5.97% for palmitic, stearic, oleic, linoleic and linolenic acid, respectively [127], which mostly coincided with fatty acid content in commercial soybean [151]. The positive outcomes of initial breeding efforts for increasing oleic and reducing linolenic acid contents at the Agricultural Institute Osijek can be seen from the significant difference between the average oleic acid content of MG 0 elite breeding lines in comparison with the MG 0 standard cultivars and significant difference between the average linolenic acid content of MG 00 and 0 elite breeding lines in comparison with the MG 00 and 0 standard cultivars, respectively (**Table 2**). Nevertheless, according to Matoša Kočar et al. [127], none of the newly developed genotypes from the Agricultural Institute Osijek exhibited desirable values for oleic or linolenic acid contents; the breeding programme would benefit from the introduction of high-oleic and low-linolenic soybean germplasm. Average year values determined in a 3-year research with RILs exhibiting a broad variation for fatty acid content ranged between 10.3 and 11.1%, 3.7 and 3.9%, 22.3 and 26.2%, 51.4 and 55% and 7.5 and 8.6% for palmitic, stearic, oleic, linoleic and linolenic acid, respectively [148]. Average fatty acid contents determined in 96 diverse accessions originating from different regions of the world and evaluated for 2 years in Brazil were 10.59, 3.21, 24.54, 52.81 and 6.36% for palmitic, stearic, oleic, linoleic and linolenic acids, respectively [152]. In Indian soybean germplasm, the average fatty acid contents were 11.15, 3.45, 25.6, 52.56 and 6.91% [126], while in Chinese they amounted to 11, 3.63, 21.47, 53.99 and 9.93% [153] for palmitic, stearic, oleic, linoleic and linolenic acids, respectively. As it can be seen from the data presented, none of the genotypes exhibited target values for oleic acid (45–60%) and linolenic acid (<3%) contents desirable for avoiding partial hydrogenation [93] which emphasises the need for further research and screening studies.

Nowadays, breeding for increased oil content can be facilitated by genetic markers or by monitoring metabolites, but although Wang et al. [103] reported that metabolic profiles of all progeny lines could be discriminated based on oil contents, there is still lack of sufficient information about the association between metabolites and oil content in soybean seeds. Nevertheless, QTL mapping and GWAS have identified 322 QTLs associated with soybean oil distributed over all 20 soybean chromosomes but mainly on chromosomes 5, 15 and 20 [97]. Fan et al. [148] identified 35 additive QTLs underlying individual fatty acid contents in a single environment and 17 additive QTLs across multiple environments or underlying multiple fatty acids. Priolli et al. [152] discovered 19 single-nucleotide polymorphism loci on 10 different chromosomes significantly associated with palmitic acid, oleic acid and total oil contents with loci and specific alleles that contributed to lower palmitic and higher oleic acid contents. Thapa et al. [154] identified independent mutations in the FAD3A associated with a reduced level of linolenic acid. Combs and Bilyeu [129] reported that seed stearic acid increased to 10–11% in lines containing combinations of FAD2-1A and FAD2-1B mutant alleles plus the SACPD-C missense mutant alleles, but this increase was associated with a decrease in the oleic acid content and did not meet the target of at least 20% stearic acid in the seed oil. Of all the reported QTLs associated with oil traits in soybean, only two (cqPro/oil-15, cqPro/oil-20) have been officially confirmed and repeatedly detected in several different populations, both associated with protein and oil contents and each showing opposite additive effect directions for the two traits, which is why identifying an environmentally stable major QTL regulating seed oil content is of crucial importance [100, 102, 103, 155]. Moreover, QTLs directly related to seed oil accumulation in soybean have not been cloned, so the underlying mechanism has not been thoroughly elucidated to date [133]. Although advance in oil content has been achieved over time, further increasing oil content by traditional breeding based on genetic crossing and phenotypic selection may be difficult and inefficient, because of the polygenic nature of oil regulation and the majority of oil-related loci having varying additive, epistatic or QTL × E effects [156–158]. Some studies suggest that increasing soybean oil could be achieved by genetic engineering of transcription factors involved in oil accumulation [159, 160]. As further increasing oil content is deemed challenging without genetic engineering [156–158] and taking into consideration that European oil industry is dominated by sunflower and rapeseed [1], it is evident that improving oil content and oil quality in Europe is not economically as important as in regions of the world where soybean is the main stock for oil production. Nevertheless, research and genotype screening for determining germplasm with favourable QTLs controlling oil traits are of great importance, especially since fatty acid phenotypes that meet the breeding objectives of providing improved oil quality are still lacking, not just in European soybean germplasm but in genotypes from all regions of the world.

4. Soluble sugars

Totally dry soybean seed contains 33% DW of carbohydrates on average, of which 16.6% DW is soluble sugars [25]. The significance of the soluble sugar profile is in its effect on quality, digestibility and nutritional value of soybean for food and feed. Five main soluble sugars in soybean seed are glucose, fructose, sucrose, raffinose and stachyose with sucrose and stachyose being the predominant ones [161]. Monosaccharides glucose and fructose and disaccharide sucrose can be easily digested and give soybean food products their characteristic sweet taste, whereas indigestible galactooligosaccharides (stachyose and raffinose) limit soybean food and feed nutritional value, reduce metabolizable energy and cause irritation of the gastrointestinal tract in humans and animals [162, 163]. In consequence, the use of soybean for food and feed is limited, but as the demand for animal feed in Europe rises [41], lowering galactooligosaccharides would increase the share of soybean seed used for feeding livestock, thus increasing the market for seed production. Furthermore, due to the increasing awareness of the health benefits connected with soy food consumption, more favourable saccharide profile would make soybean more amenable to human consumption. However, oligosaccharides are not necessarily undesirable because they serve as transport carbohydrates in the phloem, as cryoprotectants, and reportedly play a positive role in desiccation tolerance during seed maturation [164]. Furthermore, in human intestines, galactooligosaccharides ferment into low-chain fatty acids with prebiotic qualities, which make soybean seed interesting for pharmaceutical as well as functional food industry [165].

Broad-sense heritability is considered to be moderate to high for monosaccharides and high for oligosaccharides [166–168]. Geater et al. [169] determined that the differences in total sugar (TS) content among tested genotypes were constant in all environments, but significant year effect was determined for raffinose. Taira [170] found G to be the larger source of variation than E for raffinose and stachyose, whereas the opposite was true for TS and sucrose. Both G and E were statistically significant sources of variation for sucrose in the research by Maughan et al. [171]. Matoša Kočar et al. [172] reported that TS content was mostly under genetic control, whereas individual soluble sugars, as well as total oligosaccharides (TO), were significantly influenced by the year which means breeding for TS content should be much more predictable than altering sugar profiles in that particular set of genotypes. In mentioned research, glucose, fructose and raffinose contents were higher in years with higher temperatures, whereas the opposite was true for TO, sucrose and stachyose contents [172]. Wolf et al. [173] concluded that sucrose and stachyose concentrations decreased with the rise of average year temperature; however, the temperature had no effect on glucose, fructose and raffinose. The often occurring absence of G x E interaction [169], together with the determined maternal and additive effects [171–174], facilitates the selection and breeding for improved saccharide contents in soybean. To lower the expenses and reduce the time invested in analyses, plant breeders rely on favourable relationships between important traits which enable effective indirect selection. Significant positive correlation between raffinose and stachyose [161, 175] and significant and very strong negative relationship between sucrose and raffinose and/or stachyose [175, 176] enable breeders to increase the contents of favourable soluble sugars while decreasing the contents of unfavourable oligosaccharides simultaneously. Furthermore, Akond et al. [177] noted that the lack of strong correlations among tested soluble sugars indicated it should be possible to obtain lines that are high in sucrose but low in raffinose and stachyose. No significant correlation between raffinose, stachyose and sucrose was noticed by Geater et al. [169] as well. However, Hartwig et al. [178] reported a significant moderate positive correlation between raffinose and sucrose, while Hou et al. [161] determined the relationship between sucrose and raffinose to be significant, positive and strong. Another unfavourable relationship is the negative correlation determined between protein content and TS and protein content and sucrose [84, 178–179], whereas the negative correlation determined between protein content and raffinose [178] and between stachyose plus raffinose and protein content [84] is considered desirable.

The difference in correlation reports from different researchers emphasises the importance of determining trait relationships in each breeding population.

Significant difference in sugar contents, i.e. variability among soybean genotypes that justifies selection, was determined in many studies [25, 161, 172, 176, 180–182]. Geater and Fehr [181] reported TS determined in 23 soybean cultivars tested at eight Iowa locations to range between 2.19 and 18.4%. The range for TS content determined by Matoša Kočar et al. [172] in a 3-year trial (2010–2012) with 22 soybean genotypes (MG 00-II) created at the Agricultural Institute Osijek (Croatia) was 5.69–7.68% with the average value being 6.62%. In the same study, glucose content ranged 0.16–0.29%, fructose 0.11–0.28%, sucrose 2.32–3.46%, raffinose 0.66–1.10% and stachyose 1.64–2.41% [173]. Breeding for improved soluble sugar content is at its beginnings at the Agricultural Institute Osijek, but some progress

MG	Elite breeding lines		Standard cultivars		t	P
	Range	**Mean**	**Range**	**Mean**		
	Glucose (% DW)					
00	0.17–0.33	0.25	0.12–0.27	0.19	3.05^{**}	P < 0.01
0	0.11–0.35	0.23	0.13–0.19	0.16	6.58^{**}	P < 0.01
I	0.13–0.42	0.24	0.12–0.23	0.19	2.84^{**}	P < 0.01
	Fructose (% DW)					
00	0.04–0.30	0.18	0.05–0.25	0.16	0.65^{ns}	P > 0.05
0	0.02–0.38	0.17	0.02–0.18	0.11	2.87^{**}	P < 0.01
I	0.02–0.43	0.21	0.02–0.19	0.13	3.11^{**}	P < 0.01
	Sucrose (% DW)					
00	2.56–3.91	3.18	2.20–2.83	2.62	4.19^{**}	P < 0.01
0	1.77–3.96	3.01	1.89–2.71	2.32	5.74^{**}	P < 0.01
I	1.13–3.78	3.03	2.34–2.95	2.63	4.55^{**}	P < 0.01
	Total oligosaccharides (% DW)					
00	2.47–3.24	2.76	2.66–3.29	3.08	3.95^{**}	P < 0.01
0	1.81–3.53	2.75	3.02–3.34	3.18	6.83^{**}	P < 0.01
I	1.05–3.38	2.66	2.47–3.38	3.04	2.94^{**}	P < 0.01
	Total soluble sugars (% DW)					
00	5.23–7.67	6.97	5.56–6.95	6.09	3.99^{**}	P < 0.01
0	4.96–9.19	6.67	5.01–6.63	5.73	4.17^{**}	P < 0.01
I	5.04–9.21	6.68	2.95–7.25	5.69	1.75^{ns}	P > 0.05

The means were tested with Student's t-test with * and ** indicating significant differences at $P \leq 0.05$ and $P \leq 0.01$, respectively, and ns indicating no significant difference.

Table 3. The 3-year (2010–2012) soluble sugar content range and mean for different maturity group elite breeding lines created at the Agricultural Institute Osijek in comparison to standard cultivars.

has been achieved in elite breeding lines, in which higher glucose, fructose, sucrose and TS contents and a lower TO content have been determined than the standard cultivars of the respective MG (**Table 3**). Hou et al. [161] investigated worldwide soybean germplasm collections with 241 genotypes and reported average contents for TS, glucose, fructose, sucrose, raffinose and stachyose to be 96.4, 5.4, 4.4, 46.8, 8.3 and 31.7 mg g^{-1}, respectively. Values varied in wide ranges with genotypes from Peru having the least TS (72.6 mg g^{-1}) but the lowest contents of raffinose (0.5 mg g^{-1}) and stachyose (11.8 mg g^{-1}) as well [161]. The highest contents for favourable glucose (23.9 mg g^{-1}) and fructose (25.2 mg g^{-1}) were determined in genotypes from Peru as well, while the highest sucrose content (83.1 mg g^{-1}) was determined in genotypes from Angola [161]. If considering only genotypes from European countries, TS contents varied between 91.5 (Bulgaria) and 111.9 mg g^{-1} (Poland) [161]. The lowest glucose and fructose contents (0.3 mg g^{-1}) together with the highest sucrose (49.2 mg g^{-1}) and stachyose (39.2 mg g^{-1}) contents were found in a genotype from Germany, whereas the highest glucose (15.4 mg g^{-1}) and fructose (14.9 mg g^{-1}) but the lowest sucrose (16.5 mg g^{-1}), raffinose (5.2 mg g^{-1}) and stachyose (26.7 mg g^{-1}) contents were determined in three genotypes from Turkey [161]. Geater et al. [169] determined 6.2–7.1% range for sucrose, 0.49–0.58% range for raffinose and 4.7–4.9% range for stachyose in 16 small-seeded soybean cultivars. Wilcox and Shibles [84] reported combined stachyose and raffinose content in 43 random breeding lines varied in seed protein content and grown in three environments to be 38.5–42.2 g kg^{-1} and sucrose content to be 43.3–56.7 kg^{-1}. Wide content ranges determined for soybean sugar traits [25, 161, 178] imply there is sufficient variability for altering TS or individual sugar contents. Furthermore, increasing the digestibility and taste of European soybean meal or food products necessitates the introduction of cultivars with specifically tailored sugar profiles intended for human consumption to be used as parent donors in European soybean breeding programmes, which earlier mostly have not put their focus on soluble sugars.

The introduction of germplasm with favourable sugar profiles into European breeding programmes and selection of favourable phenotypes after backcrossing can be facilitated by marker-assisted selection. The first set of markers associated with sucrose content in soybean was discovered in the F2 population of an interspecific cross between *Glycine max* and *Glycine soja* detecting six QTL on linkage groups (LGs), A, E, F, I, L and M [171]. A novel allele of the putative soybean raffinose synthase gene (RS2) associated with the low raffinose and stachyose and elevated levels of sucrose content was discovered in PI200508 in Korea [183] which was used as a donor parent in crosses with three popular Korean soybean cultivars. The effects of RS2 allele were confirmed in the F2 progenies through the use of allele-specific molecular markers proving that breeding and selection for low raffinose-type soybean can be done efficiently [183]. Akond et al. [177] identified and mapped 14 significant QTLs for sucrose, raffinose and stachyose contents on eight different LGs and chromosomes with only two QTLs underlying seed sucrose and stachyose contents being previously mapped. Although previous studies identified QTL associated with seed sucrose, raffinose and stachyose contents in different populations [166, 167, 171, 173, 177], only 37

QTLs for sucrose and no QTL for raffinose and stachyose are found in SoyBase [97] to date, emphasising the necessity for more researches. Nevertheless, a high-sucrose, low-raffinose and low-stachyose germplasm was developed [168] which can be introduced to European soybean breeding programmes to be used as a parent for food-grade cultivars.

5. Isoflavones

Soybean is considered to be the most abundant natural source of isoflavones in the human and animal diet [184]. Isoflavones are the main components of flavonoids and the most common form of phytoestrogens, i.e. non-steroidal compounds with oestrogen-like biological properties [185]. They are considered nutraceuticals as it is claimed they have potential benefits in preventing the development of cardiovascular diseases and cancers, menopausal symptoms and osteoporosis, as well as antifungal and antioxidant properties [186–188]. Their role in plants is to encourage infection and nodulation by nitrogen-fixing bacteria [189] and to help in abiotic and biotic stress resistance [190]. The most studied isoflavones are genistein, daidzein and glycitein [191]. As a quantitative trait controlled by many minor genes [192], isoflavone content in soybean seed largely depends on the environment [185, 193–195], to the extent that even the smallest changes in the microclimate can cause significant changes in the isoflavone contents. For example, it is reported that lower temperatures during the seed fill period increase the isoflavone content, whereas environments which were warmer and drier resulted in lower isoflavone contents [196–199]. In 76 environments during 12 years, Carrera and Dardanelli [200] found a 91% decrease in total isoflavone content with mean temperatures during the seed development rising from 14.1 to 26.7°C. Nevertheless, Morrison et al. [201] evaluated 14 cultivars across 12 years at one location determining that high mean temperature during seed development did not result in isoflavone concentration decrease which might be due to the fact that only one location provides a somewhat narrower range of temperature variation [200]. The negative impact of water deficit on soybean isoflavone concentration was confirmed in many earlier studies [192, 198, 202, 203] as well. Furthermore, synthesis of isoflavones in soybean seed depends on the geographical origin, seed size, seed colour, maturity, disease tolerance and resistance to insects as well [185, 187, 204–206]. Nevertheless, isoflavone broad-sense heritability was estimated to range from moderate [207] to high [192], indicating that genotype effects were high enough to enable efficient improvement of isoflavone contents [194, 208]. The predominance of genotype effects over environmental was confirmed by Gutierrez-Gonzalez et al. [192], Murphy et al. [199] and Hoeck et al. [206]. Successful high-isoflavone cultivar development necessitates investigating the relationships between isoflavones and other important traits. For example, total isoflavone (TI) content was reported to be in a positive [204, 205, 209, 210] and negative [211] correlation with seed yield, but the lack of interrelationship has also been reported [199, 212], which suggests that the development of high-yield, high-isoflavone cultivars would be possible. The relationship between isoflavone content and each oil and protein content was reported to be negative [194, 199, 205, 209, 211], but no correlation between isoflavones and proteins was reported as well

[212]. Conflicting reports on the relationships between total seed isoflavone content and other traits necessitate the need for studies including a wide range of genetic material grown across a range of environments to provide additional insight into the associations [199].

Because of the many abiotic and biotic factors influencing the isoflavone content, it is expected to vary in wide ranges. Wang and Murphy [213] found TI content to vary from 1176 to 3309 μg/g within a single cultivar of soybean. Gutierrez-Gonzales et al. [192] reported of extremely wide content range (12.4–317.4 mg 100^{-1} g DM) for TI determined in the population of RILs in the USA. Murphy et al. [199] determined 160–370 mg 100^{-1} g range for TI content in RIL population tested in 2-year, multilocation trials in Canada, while Adie et al. [194] determined 14.97–39.85 mg 100^{-1} g content range for 10 soybean lines tested during 1-year trial on eight locations in Indonesia. In Europe, Cvejić et al. [208] found TI content in 20 F1 soybean progenies ranged from 156 to 366 mg 100^{-1} g, while Bursać et al. [214] determined the TI content range of 211 to 524 mg 100 g^{-1} for different seed coat-coloured offsprings derived from the single cross between commercial variety and germplasm collection genotype with black seed coat. Matoša Kočar et al. [195] investigated 22 MG 00 to II soybean genotypes from the Agricultural Institute Osijek (Croatia) during 3 years (2010–2012) and determined average TI content varied from 124.06 to 286.2 mg 100^{-1} g, while average daidzein, glycitein and genistein contents were 49.27 mg 100^{-1} g, 20.90 mg 100^{-1} g and 94.41 mg 100-1 g DM, respectively. Although breeding of isoflavone content at the Agricultural Institute Osijek is at its beginnings, some shifts in the right direction have been made which can be seen from the higher average isoflavone contents in elite breeding lines than in standard cultivars of the same MG (**Table 4**). In Matoša Kočar et al. [195] study, genistein was the most abundant isoflavone component, followed by daidzein and glycitein. The same order of abundance was determined in the research by Lozovaya et al. [198] for two French and three US cultivars, whereas Sumardi et al. [215] researched 34 black soybeans in Indonesia and found that daidzein content was higher in 31 genotypes. In the research done by Cvejić et al. [208], Bursać et al. [214] and Tepavčević et al. [216], total daidzein was the highest followed by total genistein and total glycitein. Gutierrez-Gonzalez et al. [192] found that glycitein was the most abundant isoflavone, followed by genistein and daidzein. Although all three isoflavones are considered to have health benefits, the order of abundance is important because genistein is reported to have approximately 10 times higher biological activity compared to daidzein and glycitein [217], so genotypes with high genistein content should be favoured in selection processes aiming to create soybean genotypes suitable for food and dietetic supplement industries.

The large influence the environmental factors have on the isoflavone content emphasises the necessity for molecular research and identification of molecular markers associated with favourable isoflavone profiles. To date, almost 90 QTLs associated with seed isoflavone content are registered in the SoyBase [97]. Akond et al. [193] identified 16 QTLs for seed isoflavones content on 12 different chromosomes (Chr) or linkage groups. Wang et al. [218] identified 33 expression QTLs on 15 soybean chromosomes, with 5 of them overlapping with phenotype QTL. In the overlapping region, 11 candidate genes underlying the accumulation of isoflavones were discovered which could be beneficial for the development of marker-assisted selection to breed soybean cultivars with high-isoflavone contents [218]. Furthermore, Wang et al. [219] identified 23 new isoflavone content QTLs and 34 QTLs in total in 130 RILs derived

MG	Elite breeding lines		Standard cultivars		t	P
	Range	Mean	Range	Mean		
	Daidzein (mg 100 g^{-1})					
00	10.65–106.48	65.07	20.34–63.99	45.02	2.11*	P < 0.05
0	11.08–93.39	50.54	19.02–48.53	36.71	3.28**	P < 0.01
I	25.03–91.35	51.26	9.81–53.51	36.01	2.70**	P < 0.01
	Glycitein (mg 100 g^{-1})					
00	10.72–40.31	24.26	19.08–27.24	22.77	0.68^{ns}	P > 0.05
0	13.09–41.76	21.78	12.97–18.62	15.54	6.46**	P < 0.01
I	10.09–47.38	19.91	9.70–18.94	13.41	4.71**	P < 0.01
	Genistein (mg 100 g^{-1})					
00	43.35–355.31	184.85	57.59–128.23	93.89	4.08**	P < 0.01
0	52.32–112.25	80.49	54.82–86.38	67.44	2.86**	P < 0.01
I	48.19–142.22	92.24	54.48–104.39	83.80	1.12^{ns}	P > 0.05
	Total isoflavones (mg 100 g^{-1})					
00	70.49–501.79	273.18	97.22–206.06	161.68	3.50**	P < 0.01
0	87.61–212.99	150.05	115.12–148.60	136.25	0.76^{ns}	P > 0.05
I	91.34–239.75	160.75	71.14–179.53	129.94	2.34*	P < 0.05

The means were tested with Student's t-test with * and ** indicating significant differences at P ≤ 0.05 and P ≤ 0.01, respectively, and ns indicating no significant difference.

Table 4. The 3-year (2010–2012) isoflavone content range and mean for different maturity group elite breeding lines created at the agricultural institute Osijek in comparison to standard cultivars.

from the cross between high-isoflavone and low-isoflavone cultivars of which 6, 7, 10 and 11 QTLs were associated with daidzein, glycitein, genistein and TI, respectively. Akond et al. [193] identified three QTLs on three different linkage groups, one controlling daidzein content and two controlling glycitein content. Furthermore, QTL epistatic interactions are also thought to contribute to isoflavone variation [205]. Although many QTL have been discovered for isoflavone content, these are mainly minor effect often influenced by the environment, so discovering sufficient loci associated with isoflavone content stable in different environments that are going to significantly aid in selection is yet to be accomplished [219].

6. Conclusion

Although genetic enhancement of soybean seed quality contributes to advances in processing industries and improves the added value properties of final soybean products, soybean as a commodity is still mostly being paid for by weight and not by composition. Considering the fact that in Europe soybean is predominantly used as animal feed, i.e. as a protein source rather than oilseed crop, it is expected of European breeders to mainly focus on increasing the seed yield and protein content, since this would be most profitable for soybean producers. As soybean is becoming more appealing for human consumption in Europe, because of high nutritional value and health-promoting traits, breeding for improved amino acid, fatty acid and soluble sugar compositions as well as for increased isoflavone content is gaining importance but still not on a larger scale. Nevertheless, all efforts in describing the variability of important traits through research are invaluable not only for creating superior progeny but for germplasm preservation to oppose the narrowing of the genetic base. Although soybean is widely adaptable, significant environment effects for different seed quality traits, as well as the increased frequency of adverse weather events, emphasise the need for the development of local cultivars with improved performance and stability. The need for creating European soybean cultivars emanates from the demand for non-GM soybean as well, which motivates European breeders to use conventional breeding methods focusing on phenotype selection and MAS. Positive outcomes of breeding for improved seed quality recorded at the Agricultural Institute Osijek (Croatia) and other European soybean breeding programmes indicate that progress can be achieved even without genetic engineering. As this paper mainly focuses on MG 00 to II, commonly sown in Central and South-eastern Europe but suitable for growing in almost all European regions, information presented should be useful for soybean breeders and researches all over Europe and could promote the exchange of germplasm for introducing diversity, which is a prerequisite for any genetic advance.

Author details

Aleksandra Sudarić[1,2], Maja Matoša Kočar[1*], Tomislav Duvnjak[1], Zvonimir Zdunić[1,2] and Antonela Markulj Kulundžić[1]

*Address all correspondence to: maja.matosa@poljinos.hr

1 Agricultural Institute Osijek, Osijek, Republic of Croatia

2 Centre of Excellence for Biodiversity and Molecular Plant Breeding, Faculty of Agriculture, University of Zagreb, Zagreb, Republic of Croatia

References

[1] FAOSTAT Database [Internet]. 2019. Available from: http://faostat3.fao.org/home/E [Accessed: 24 July 2019]

[2] Cober ER, Cianzio SR, Pantalone VR, Rajcan I. Soybean. In: Vollmann J, Rajcan I, editors. Oil Crops. Handbook of Plant Breeding. Vol. 4. New York, USA: Springer Nature Switzerland AG; 2009. pp. 57-90

[3] Tsuda T, Osawa T, Ohshima K, Kawakishi S. Antioxidative pigments isolated from the seeds of *Phaseolus vulgaris* L. Journal of Agricultural and Food Chemistry. 1994;**42**:248-251

[4] Zou P. Traditional Chinese medicine, food therapy, and hypertension control: A narrative review of Chinese literature. The American Journal of Chinese Medicine. 2016;**44**:1-16. DOI: 10.1142/S0192415X16500889

[5] Pascal T, Rodriguez Cerezo E. Markets for Non-Genetically Modified, Identity-Preserved Soybean in the EU. JRC Science and Policy Report. Luxembourg: Publications Office of the European Union; 2015. DOI: 10.2791/949110

[6] Dima DC. Soybean demonstration platforms: The bond between breeding, technology and farming in central and Eastern Europe. Agriculture and Agricultural Science Procedia. 2016;**10**:10-17. DOI: 10.1016/j.aaspro.2016.09.003

[7] Kurasch AK, Hahn V, Leiser WL, Vollmann J, Schori A, Bétrix CA, et al. Identification of mega-environments in Europe and effect of allelic variation at maturity *E. loci* on adaptation of European soybean. Plant, Cell & Environment. 2017;**40**:765-778

[8] Jug D, Jug I, Brozović B, Vukadinović V, Stipešević B, Đurđević B. The role of conservation agriculture in mitigation and adaptation to climate change. Poljoprivreda. 2018;**24**(1):35-44. DOI: 10.18047/poljo.24.1.5

[9] Li G, Ra WH, Park JW, Kwon SW, Lee JH, Park CB, et al. Developing EST-SSR markers to study molecular diversity in Liriope and Ophiopogon. Biochemical Systematics and Ecology. 2011;**39**:241-252. DOI: 10.1016/j.bse.2011.08.012

[10] Vratarić M, Sudarić A. Soja. Osijek: Poljoprivredni Institut Osijek; 2008. pp. 1-459

[11] Burton JW. Soybean (*Glycine max* (L.) Merr.). Field Crops Research. 1997;**53**:171-186

[12] Gizlice Z, Carter TE Jr, Gerig TM, Burton JW. Genetic diversity patterns in North American public soybean cultivars based on coefficient of parentage. Crop Science. 1996;**33**:753-765. DOI: 10.1590/S1415-47572006000400019

[13] Mudibu J, Nkongolo KKC, Kalonji-Mbuyi A. Morphovariability and agronomic characteristics of soybean accessions from the Democratic Republic of Congo (DR-Congo) gene pool. Journal of Plant Breeding and Crop Science. 2011;**3**:660-668

[14] Malik MFA, Qureshi A, Ashraf M, Khan M, Javed A. Evaluation of genetic diversity in soybean (*Glycine max*) lines using seed protein electrophoresis. Australian Journal of Crop Science. 2009;**3**:107-112

[15] Salimi S. Relationships of some soybean genotypes based on morphological and biochemical markers. International Agronomy and Plant Production. 2013;**4**:2237-2243

[16] Mulato BM, Möller M, Zucchi MI, Quecini V, Pinheiro JB. Genetic diversity in soybean germplasm identified by SSR and EST-SSR markers. Pesquisa Agropecuária Brasileira. 2010;**45**:276-283

[17] Tantasawat P, Trongchuen J, Prajongjai T, Jenweerawat S, Chaowiset W. SSR analysis of soybean (*Glycine max* (L.) Merr.) genetic relationship and variety identification in Thailand. Australian Journal of Crop Science. 2011;**5**:283-290

[18] Moose SP, Mumm RH. Molecular plant breeding as the foundation for 21st century crop improvement. Plant Physiology. 2008;**147**:969-977. DOI: 10.1104/pp.108.118232

[19] Harlan JR. Genetics of disaster. Journal of Environmental Quality. 1972;**1**:212-215

[20] Zhang G, Xu S, Mao W, Hu Q, Gong Y. Determination of the genetic diversity of vegetable soybean (*Glycine max* (L.) Merr.) using EST-SSR markers. Journal of Zhejiang University – Science B (Biomedicine & Biotechnology). 2013;**14**:279-288. DOI: 10.1631/jzus.B1200243

[21] Perić V, Nikolić A, Babić V, Sudarić A, Srebrić M, Đorđević V, et al. Genetic relatedness of soybean genotypes. Genetika-Belgrade. 2014;**46**:839-854. DOI: 10.2298/GENSR1403839P

[22] Hahn V, Würschum T. Molecular genetic characterization of Central European soybean breeding germplasm. Plant Breeding. 2014;**133**:748-755. DOI: 10.1111/pbr.12212

[23] Grainger CM, Rajcan I. Characterization of the genetic changes in a multi-generational pedigree of an elite Canadian soybean cultivar. Theoretical and Applied Genetics. 2014;**127**:211-229. DOI: 10.1007/s00122-013-2211-9

[24] Varshney RK, Thudi M, Pandey MK, Tardieu F, Ojiewo C, Vadez V, et al. Accelerating genetic gains in legumes for the development of prosperous smallholder agriculture: Integrating genomics, phenotyping, systems modelling and agronomy. Journal of Experimental Botany. 2018;**69**:3293-3312. DOI: 10.1093/jxb/ery088

[25] Hymowitz T, Collins FI. Variability of sugar content of seed of *Glycine max* (L.) Merr. and *G. soja* Serb. and Zucco. Agronomy Journal. 1974;**66**:239-240. DOI: 10.2134/agronj1974.00021962006600020017x

[26] Cromwell GL. Soybean Meal: An Exceptional Protein Source. Ankeny, IA: Soybean Meal InfoCenter [Internet]. 2012. Available from: https://www.soymeal.org/wpcontent/uploads/2018/04/soybean_meal_an_exceptional_protein_source.pdf [Accessed: 25 July 2019]

[27] Hasler CM. The cardiovascular effects of soy products. Journal of Cardiovascular Nursing. 2002;**16**:50-63

[28] Zarkadas CG, Voldeng HD, Yu ZR, Choi V. Assessment of the protein quality of nine northern adapted yellow and brown seed coated soybean cultivars by amino acid analysis. Journal of Agricultural and Food Chemistry. 1999;**47**:5009-5018. DOI: 10.1021/jf981381r

[29] Everson G, Heckert A. The biological value of some leguminous sources of protein. Journal of the American Dietetic Association. 1944;**20**:81-82

[30] Wagner JR, Sorgentini DA, Anon MC. Thermal and electrophoretic behavior, hydrophobicity, and some functional properties of acid treated soy isolates. Journal of Agricultural and Food Chemistry. 1996;**44**:1881-1889. DOI: 10.1021/jf950444s

[31] Cai T, Chang KC. Processing effect on soybean storage proteins and their relationship with tofu quality. Journal of Agricultural and Food Chemistry. 1999;**47**:720-727. DOI: 10.1021/jf980571z

[32] Žilić SM, Barać MB, Pešić MB, Mladenović Drinić SD, Ignjatović-Micić DD, Srebrić MB. Characterization of proteins from kernel of different soybean varieties. Journal of the Science of Food and Agriculture. 2010;**91**:60-67. DOI: 10.1002/jsfa.4148

[33] Kito M, Moriyama T, Kimura Y, Kambara H. Changes in plasma lipid levels in young healthy volunteers by adding an extruder-cooked soy protein to conventional meals. Bioscience, Biotechnology, and Biochemistry. 1993;**57**:354-355. DOI: 10.1271/bbb.57.354

[34] Poysa V, Woodrow L, Yu K. Effect of soy protein subunit composition on tofu quality. Food Research International. 2006;**39**:309-317. DOI: 10.1016/j.foodres.2005.08.003

[35] Liener IE. Implications of antinutritional components in soybean foods. Critical Reviews in Food Science and Nutrition. 1994;**34**:31-67. DOI: 10.1080/10408399409527649

[36] Yin Y, Fatufe AA, Blachier F. Soya bean meal and its extensive use in livestock feeding and nutrition, soybean and nutrition. In: El-Shemy H, editor. Soybean and Nutrition. Rijeka, Croatia: InTech; 2011. pp. 369-384. DOI: 10.5772/18823

[37] Yasothai R. Antinutritional factors in soybean meal and its deactivation. International Journal of Science, Environment and Technology. 2016;**5**:3793-3797

[38] SoyStats [Internet]. 2019. Available from: http://soystats.com/international-world-protein-meal-consumption/ [Accessed: July 30, 2019]

[39] Gatrell S, Lum K, Kim J, Lei XG. Nonruminant nutrition symposium: Potential of defatted microalgae from the biofuel industry as an ingredient to replace corn and soybean meal in swine and poultry diets. Journal of Animal Science. 2014;**92**:1306-1314. DOI: 10.2527/jas.2013-7250

[40] Jobbágy EG, Sala OE. The imprint of crop choice on global nutrient needs. Environmental Research Letters. 2014;**9**:084014. DOI: 10.1088/1748-9326/9/8/084014

[41] Bues A, Preißel S, Reckling M, Kuhlman T, Topp K, Watson C, et al. The Environmental Role of Protein Crops in the New Common Agricultural Policy [Internet]. 2013. Available from: http://www.europarl.europa.eu/RegData/etudes/etudes/join/2013/495856/IPOL-AGRI_ET(2013)495856_EN.pdf [Accessed: 14 July 2019]

[42] Alexandratos N, Bruinsma J. World agriculture towards 2030/2050: The 2012 revision. ESA Working Paper No. 12-03. Rome, FAO [Internet]. 2012. Available from: http://www.fao.org/3/a-ap106e.pdf [Accessed: 20 July 2019]

[43] Mottet A, de Haan C, Falcucci A, Tempio G, Opio C, Gerber P. Livestock: On our plates or eating at our table? A new analysis of the feed/food debate. Global Food Security. 2017;**14**:1-8

[44] Watson CA, Reckling M, Preissel S, Bachinger J, Bergkvist G, Kuhlman T, et al. Grain legume production and use in European agricultural systems. Advances in Agronomy. 2017;**144**:235-303. DOI: 10.1016/bs.agron.2017.03.003

[45] Carbon Trust. Carbon-Footprinting Software: Footprint Expert. London: Carbon Trust [Internet]. 2013. Available from: http://www.carbontrust.com/software [Accessed: 16 July 2019]

[46] Chéreau D, Videcoq P, Ruffieux C, Pichon L, Motte JC, Belaid S, et al. Combination of existing and alternative technologies to promote oilseeds and pulses proteins in food applications. Oilseeds and Fats, Crops and Lipids. 2016;**23**:D406. DOI: 10.1051/ocl/2016020

[47] Flachowsky G, Meyer U, Südekum KH. Land use for edible protein of animal origin—A review. Animals. 2017;**7**:25. DOI: 10.3390/ani7030025.

[48] Zander P, Amjath-Babu TS, Preissel S, Reckling M, Bues A, Schläfke N, et al. Grain legume decline and potential recovery in European agriculture: A review. Agronomy for Sustainable Development. 2016;**36**:26. DOI: 10.1007/s13593-016-0365-y

[49] Roman GV, Epure LI, Toader M, Lombardi AR. Grain legumes—Main source of vegetal proteins for European consumption. AgroLife Scientific Journal. 2016;**1**:178-183

[50] Chetana C, Rusub T, Chetana F, Simona A. Influence of soil tillage systems and weed control treatments on root nodules, production and qualitative indicators of soybean. Procedia Technology. 2016;**22**:457-464. DOI: /10.1016/j.protcy.2016.01.088

[51] Latifi N. Yield and morphological response of soybean to time of irrigation and sowing rate. Dissertation Abstracts International. 1989;**40**:5088-5098

[52] Kirnak H, Dogan E, Turkoglu H. Effect of drip irrigation intensity on soybean seed yield and quality in the semi-arid Harran plain, Turkey. Spanish Journal of Agricultural Research. 2010;**8**:1208-1217. DOI: 10.5424/sjar/2010084-1239

[53] Ghassemi-Golezani K, Farshbaf-Jafari S. Influence of water deficit on oil and protein accumulation in soybean grains. International Journal of Plant, Animal and Environmental Sciences. 2012;**2**:46-52

[54] Foroud N, MuÈndel HH, Saindon G, Entz T. Effect of level and timing of moisture stress on soybean yield, protein, and oil responses. Field Crops Research. 1993;**31**:195-209. DOI: 10.1016/0378-4290(93)90062-R

[55] Bouniols A, Texier V, Mondies M, Piva G. Soybean seed quality among genotypes and crop management: Field experiment and simulation. Eurosoya. 1997;**11**:87-99

[56] Afza R, Hardarson G, Zapata F, Danso SKA. Effects of delayed soil and foliar N fertilisation on yield and N2 fixation of soybean. Plant and Soil. 1987;**97**:361-368

[57] Isfan D. Fertilizer nitrogen uptake by soybean as related to cultivars and time of application using 15N technique. Journal of Plant Nutrition. 1991;**14**:1369-1380. DOI: 10.1080/01904169109364292

[58] Paek NC, Imsande J, Shoemaker RC, Shibles R. Nutritional control of soybean seed storage protein. Crop Science. 1997;**37**:498-503. DOI: 10.2135/cropsci1997.0011183X003700020031x

[59] Grabau LJ, Blevins DG, Minor HC. Stem infusions enhanced methionine content of soybean storage protein. Plant Physiology. 1986;**82**:1013-1018. DOI: 10.1104/pp.82.4.1013

[60] Zimmer S, Messmer M, Haase T, Piepho HP, Mindermann A, Schulz H, et al. Effects of soybean variety and Bradyrhizobium strains on yield, protein content and biological nitrogen fixation under cool growing conditions in Germany. European Journal of Agronomy. 2016;**72**:38-46. DOI: 10.1016/j.eja.2015.09.008

[61] Vollmann J, Fritz CN, Wagentristl H, Ruckenbauer P. Environmental and genetic variation of soybean seed protein content under central European growing conditions. Journal of the Science of Food and Agriculture. 2000;**80**:1300-1306. DOI: 10.1002/1097-0010(200007)80:9<1300:AID-JSFA640>3.3.CO;2-9

[62] Kim SW, Less JF, Wang L, Yan T, Kiron V, Kaushik SJ, et al. Meeting global feed protein demand: Challenge, opportunity, and strategy. Annual Review of Animal Biosciences. 2019;**7**:221-243. DOI: 10.1146/annurev-animal-030117-014838

[63] Sudarić A, Šimić D, Vratarić M. Characterization of genotype by environment interactions in soybean breeding programmes of Southeast Europe. Plant Breeding. 2006;**125**: 191-194. DOI: 10.1111/j.1439-0523.2006.01185.x

[64] Bueno RD, Borges LL, Arruda KMA, Bhering LL, de Barros EG, Moreira MA. Genetic parameters and genotype x environment interaction for productivity, oil and protein content in soybean. African Journal of Agricultural Research. 2013;**8**:4853-4859. DOI: 10.5897/AJAR2013.6924

[65] Ghodrati G. Study of genetic variation and broad sense heritability for some qualitative and quantitative traits in soybean (*Glycine max* L.) genotype. Current Opinion in Agriculture. 2013;**2**:31-35

[66] Josipović M, Sudarić A, Sudar R, Plavšić H, Marković M, Jug D, Stojić B. Influence of irrigation and variety on the soybean grain yield and quality in the no nitrogen fertilisation soil condition. In: Soil and Crop Management: Adaptation and Mitigation of Climate Change; 26-28 September 2013; Osijek. Osijek: Grafika d.o.o; 2013. pp. 237-245

[67] Rodrigues JID, Arruda KMA, Cruz CD, Piovesan ND, de Barros EG, Moreira MA. Biometric analysis of protein and oil contents of soybean genotypes in different environments. Pesquisa Agropecuária Brasileira. 2014;**49**:475-482. DOI: 10.1590/S0100-204X2014000600009

[68] Gravois KA, Bernhardt JL. Heritability x environment interactions for discoloured rice kernels. Crop Science. 2000;**40**:314-318. DOI: 10.2135/cropsci2000.402314x

[69] Malik MFA, Ashraf M, Qureshi AS, Ghafoor A. Utilisation of diverse germplasm for soybean yield improvement. Asian Journal of Plant Sciences. 2006;**5**:663-667. DOI: 10.3923/ajps.2006.663.667

[70] Kurasch AK, Hahn V, Leiser WL, Starck N, Würschum T. Phenotypic analysis of major agronomic traits in 1008 RILs from a diallel of early European soybean varieties. Crop Science. 2017;**57**:726-738. DOI: 10.2135/cropsci2016.05.0318

[71] Popović V, Vidić M, Tatić M, Jakšić S, Kostić M. Uticaj sorte i godine na prinos i komponente kvaliteta soje. Ratarstvo i Povrtarstvo. 2012;**49**:132-139. DOI: 10.5937/ratpov49-1140

[72] Hurburgh CR Jr. Long-term soybean composition patterns and their effect on processing. Journal of the American Oil Chemists' Society. 1994;**71**:1425-1427

[73] Nian H, Wang JL, Yang QK. Effects of ecological conditions of the northeast of China on protein, oil and protein+oil content in soybean seeds. Journal of Northeast Agricultural University (English Edition). 1996;**3**:1-6

[74] Zhang F, Lynch DH, Smith DL. Impact of low root temperatures in soybean [*Glycine max* (L.) Merr.] on nodulation and nitrogen fixation. Environmental and Experimental Botany. 1995;**35**:279-285. DOI: 10.1016/0098-8472(95)00017-7

[75] Gibson LR, Mullen RE. Soybean seed composition under high day and night growth temperatures. Journal of the American Oil Chemists' Society. 1996;**73**:733-737

[76] Dornbos DL Jr, Mullen RE. Soybean seed protein and oil contents and fatty acid composition adjustments by drought and temperature. Journal of the American Oil Chemists' Society. 1992;**69**:228-231

[77] Matoša Kočar M, Sudarić A, Vila S, Petrović S, Rebekić A, Josipović A, et al. Varijabilnost fenotipske ekspresije svojstava kvalitete zrna elitnih linija soje. Poljoprivreda. 2017;**23**:40-48. DOI: 10.18047/poljo.23.1.7

[78] Brim CA, Burton JW. Recurrent selection in soybeans. II. Selection for increased percent protein in seeds. Crop Science. 1979;**19**:494-498

[79] Leffel RC, Rhodes WK. Agronomic performance and economic value of high-seed protein soybean. Journal of Production Agriculture. 1993;**6**:365-368

[80] Cober ER, Voldeng HD. Developing high-protein, high-yield soybean populations and lines. Crop Science. 2000;**40**:39-42. DOI: 10.2135/cropsci2000.40139x

[81] Pannecoucque J, Goormachtigh S, Heungens K, Vleugels T, Ceusters J, Van Waes C, et al. Screening for soybean varieties suited to Belgian growing conditions based on maturity, yield components and resistance to *Sclerotinia sclerotiorum* and *Rhizoctonia solani* anastomosis group 2-2IIIB. The Journal of Agricultural Science. 2018;**156**:342-349. DOI: 10.1017/S0021859618000333

[82] Chung J, Babka HL, Graef GL, Staswick PE, Lee DJ, Cregan PB, et al. The seed protein, oil, and yield QTL on soybean linkage group I. Crop Science. 2003;**43**:1053-1067. DOI: 10.2135/cropsci2003.1053

[83] Clemente TE, Cahoon EB. Soybean oil: Genetic approaches for modification of functionality and total content. Plant Physiology. 2009;**151**:1030-1040. DOI: 10.1104/pp.109.146282

[84] Wilcox JR, Shibles RM. Interrelationships among seed quality attributes in soybean. Crop Science. 2001;**41**:11-14. DOI: 10.2135/cropsci2001.41111x

[85] Burton J, Purcell A, Walter W. Methionine concentration in soybean protein from populations selected for increased percent protein. Crop Science. 1982;**22**:430-432. DOI: 10.2135/cropsci1982.0011183X002200020054x

[86] Krober OA, Cartter JL. Relation of methionine to protein levels in soybeans. Cereal Chemistry. 1966;**43**:320-325

[87] Fogelberg F, Recknagel J. Developing soy production in central and northern Europe. In: Murphy-Bokern D, Stoddard FL, Watson CA, editors. Legumes in Cropping Systems. CAB International; 2017. pp. 109-124. DOI: 10.1079/9781780644981.0109

[88] Sudarić A, Vratarić M. Variability and interrelationships of grain quantity and quality characteristics in soybean. Die Bodenkultur. 2002;**53**:137-142

[89] Vratarić M, Sudarić A, Sudar R, Duvnjak T, Jurković D, Jurković Z. Genetic advance in quantitative traits of soybean lines within different maturity groups. Poljoprivreda. 2005;**11**:5-10

[90] Sato T, van Schoote M, Wagentristl H, Vollmann J. Effects of divergent selection for seed protein content in high-protein vs. food-grade populations of early maturity soybean. Plant Breeding. 2014;**133**:74-79. DOI: 10.1111/pbr.12138

[91] Sharma S, Kaur M, Goyal R, Gill BS. Physical characteristics and nutritional composition of some new soybean (*Glycine max* (L.) Merrill) genotypes. Journal of Food Science and Technology. 2014;**51**:551-557. DOI: 10.1007/s13197-011-0517-7

[92] Ramteke R, Kumar V, Muralidharan P, Agarwal DK. Study on genetic variability and traits interrelationship among released soybean varieties of India (*Glycine max* (L.) Merrill). Electronic Journal of Plant Breeding. 2010;**1**:1483-1487

[93] Wilson RF. Seed composition. In: Boerma HR, Specht JE, editors. Soybeans: Improvement, Production and Uses, Ed 3. Madison, Wisconsin, USA: American Society of Agronomy, Crop Science Society of America, Soil Science Society of America; 2004. pp. 621-677

[94] Thorne JC, Fehr WR. Incorporation of high-protein, exotic germplasm into soybean populations by 2- and 3-way crosses. Crop Science. 1970;**10**:652-655

[95] Wehrmann VK, Fehr WR, Cianzio SR, Cavins JF. Transfer of high seed protein to high-yielding soybean cultivars. Crop Science. 1987;**27**:927-931. DOI: 10.2135/cropsci1987.0011183X002700050020x

[96] Wilcox JR, Cavins JF. Backcrossing high seed protein to a soybean cultivar. Crop Science. 1995;**35**:1036-1041. DOI: 10.2135/cropsci1995.0011183X003500040019x

[97] SoyBase [Internet]. 2019. Available from: https://www.soybase.org/ [Accessed: 23 July 2019]

[98] Brummer EC, Graef GL, Orf J, Wilcox JR, Shoemaker R. Mapping QTLs for seed protein and oil content in eight soy-bean populations. Crop Science. 1997;**37**:370-378

[99] Sebolt AM, Shoemaker RC, Diers BW. Analysis of a quantitative trait locus allele from wild soybean that increases seed protein concentration in soybean. Crop Science. 2000;**40**:1438-1444. DOI: 10.2135/cropsci2000.4051438x

[100] Hwang EY, Song QJ, Jia GF, Specht JE, Hyten DL, Costa J, et al. A genome-wide association study of seed protein and oil content in soybean. BMC Genomics. 2014;**15**:1-12. DOI: 10.1186/1471-2164-15-1

[101] Nichols DM, Glover KD, Carlson SR, Specht JE, Diers BW. Fine mapping of a seed protein QTL on soybean linkage group I and its correlated effects on agronomic traits. Crop Science. 2006;**46**:834-839. DOI: 10.2135/cropsci205.05-0168

[102] Bandillo N, Jarquin D, Song Q, Nelson R, Cregan P, Specht J, et al. A population structure and genome-wide association analysis on the USDA soybean germplasm collection. Plant Genome. 2015;**8**:1-13. DOI: 10.3835/plantgenome2015.04.0024

[103] Wang J, Zhou P, Shi X, Yang N, Yan L, Zhao Q, et al. Primary metabolite contents are correlated with seed protein and oil traits in near-isogenic lines of soybean. The Crop Journal. DOI: 10.1016/j.cj.2019.04.002

[104] Durham D. The United Soybean Board's better bean initiative: Building United States soybean competitiveness from the inside out. The Journal of Agrobiotechnology Management & Economics. 2003;**6**:23-26

[105] Clarke E, Wiseman J. Developments in plant breeding for improved nutritional quality of soya beans. The Journal of Agricultural Science. 2000;**134**:111-124. DOI: 10.1017/S0021859699007431

[106] Fallen BD, Hatcher CN, Allen FL, Kopsell DA, Saxton AM, Chen P, et al. Soybean seed amino acid content QTL detected using the universal soy linkage panel 1.0 with 1,536 SNPs. Journal of Plant Genome Sciences. 2013;**1**:68-79. DOI: 10.5147/jpgs.2013.0089

[107] Wang XZ, Jiang GL, Song QJ, Cregan PB, Scott RA, Zhang JP, et al. Quantitative trait locus analysis of seed sulfur-containing amino acids in two recombinant inbred line populations of soybean. Euphytica. 2015;**201**:293-305. DOI: 10.1007/s10681-014-1223-0

[108] Vaughn JN, Nelson RL, Song Q, Cregan PB, Li Z, Vaughn JN, et al. The genetic architecture of seed composition in soybean is refined by genome-wide association scans across multiple populations. G3 (Bethesda). 2014;**4**:2283-2294. DOI: 10.1534/g3.114.013433

[109] Qiu H, Hao W, Gao S, Ma X, Zheng Y, Meng F, et al. Gene mining of sulfur-containing amino acid metabolic enzymes in soybean. Yi Chuan = Hereditas. 2014;**36**:934-942. DOI: 10.3724/SP.J.1005.2014.0934

[110] Panthee D, Kwanyuen P, Sams C, West D, Saxton A, Pantalone V. Quantitative trait loci for β-conglycinin (7S) and glycinin (11S) fractions of soybean storage protein. Journal of the American Oil Chemists' Society. 2004;**81**:1005-1012

[111] Ma Y, Kan G, Zhang X, Wang Y, Zhang W, Du H, et al. Quantitative trait loci (QTL) mapping for glycinin and β-conglycinin contents in soybean (*Glycine max* L. Merr.). Journal of Agricultural and Food Chemistry. 2016;**64**:3473-3483. DOI: 10.1021/acs.jafc.6b00167

[112] Boehm JD Jr, Nguyen V, Tashiro RM, Anderson D, Shi C, Wu X, et al. Genetic mapping and validation of the loci controlling 7S a′ and 11S A-type storage protein subunits in soybean [*Glycine max* (L.) Merr.]. Theoretical and Applied Genetics. 2018;**131**:659-671. DOI: 10.1007/s00122-017-3027-9

[113] Murphy PA, Resurreccion AP. Varietal and environmental differences in soybean glycinin and β-conglycinin content. Journal of Agricultural and Food Chemistry. 1984;**32**:911-915

[114] Mujoo R, Trinh DT, Perry KW. Characterization of storage proteins in different soybean varieties and their relationship to tofu yield and texture. Food Chemistry. 2003;**82**:265-273. DOI: 10.1016/S0308-8146(02)00547-2

[115] Zhang D, Lü H, Chu S, Zhang H, Zhang H, Yang Y, et al. The genetic architecture of water-soluble protein content and its genetic relationship to total protein content in soybean. Scientific Reports. 2017;**7**:5053. DOI: 10.1038/s41598-017-04685-7

[116] Krishnana HB, Jezb JM. Review: The promise and limits for enhancing sulfur-containing amino acid content of soybean seed. Plant Science. 2018;**272**:14-21. DOI: 10.1016/j.plantsci.2018.03.030

[117] SoyStats [Internet]. 2018. Available from: http://soystats.com/ [Accessed: 26 July 2019]

[118] Snyder CL, Yurchenko OP, Siloto RM, Chen X, Liu Q, Mietkiewska E, et al. Acyltransferase action in the modification of seed oil biosynthesis. New Biotechnology. 2009;**26**:11-16. DOI: 10.1016/j.nbt.2009.05.005

[119] Fehr WR. Breeding for modified fatty acid composition in soybean. Crop Science. 2007;**47**:S-72-S-87. DOI: 10.2135/cropsci2007.04.0004IPBS

[120] Baum SJ, Kris-Etherton PM, Willett WC, Lichtenstein AH, Rudel LL, Maki KC, et al. Fatty acids in cardiovascular health and disease: A comprehensive update. Journal of Clinical Lipidology. 2012;**6**:216-234. DOI: 10.1016/j.jacl.2012.04.077

[121] Crupkin M, Zambelli A. Detrimental impact of trans fats on human health: Stearic acid-rich fats as possible substitutes. Comprehensive Reviews in Food Science and Food Safety. 2008;**7**:271-279. DOI: 10.1111/j.1541-4337.2008.00045.x

[122] Jeong JE, Krishnanand KP, Chang JH, Ha BK, Kang ST, Bilyeu K, et al. A novel allele of GmSACPD-C associated with high seed stearic acid concentration in an EMS-induced mutant PE980 in soybean. Crop Science. 2018;**58**:192-203. DOI: 10.2135/cropsci2017.05.0313

[123] Johnson S, Saikia N. Fatty Acids Profile of Edible Oils and Fats in India. New Delhi, India: Centre for Science and Environment; 2009

[124] Lee GJ, Wu X, Shannon JG, Sleper DA, Nguyen HT. Soybean. In: Kole C, editor. Genome Mapping and Molecular Breeding in Plants. Berlin: Springer; 2007. pp. 1-3

[125] Duh PD, Yen WJ, Yen GC. Oxidative stability of polyunsaturated fatty acids and soybean oil in an aqueous solution with emulsifiers. Journal of the American Oil Chemists' Society. 1999;**76**:201-204

[126] Rani A, Kumar V, Verma SK, Shakya AK, Hussain SM, Chauhan GS. Interrelationship between oil content and fatty acid composition in Indian soybean (*Glycine max*) cultivars. The Indian Journal of Agricultural Sciences. 2007;**77**:59-62

[127] Matoša Kočar M, Sudarić A, Sudar R, Duvnjak T, Zdunić Z. Screening of early maturing soybean genotypes for production of high quality edible oil. Zemdirbyste-Agriculture. 2018;**105**:55-62. DOI: 10.13080/z-a.2018.105.008

[128] Brouwer IA, Wanders AJ, Katan MB. Effect of animal and industrial trans fatty acids on HDL and LDL cholesterol levels in humans—A quantitative review. PLoS One. 2010;**5**:e9434. DOI: 10.1371/journal.pone.0009434

[129] Combs R, Bilyeu K. Novel alleles of FAD2-1A induce high levels of oleic acid in soybean oil. Molecular Breeding. 2019;**39**:79. DOI: 10.1007/s11032-019-0972-9

[130] Williams CD, Whitley BM, Hoyo C, Grant DJ, Iraggi JD, Newman KA, et al. Freedland SJ. A high ratio of dietary n-6/n-3 polyunsaturated fatty acids is associated with increased risk of prostate cancer. Nutrition Research. 2011;**31**:1-8. DOI: 10.1016/j.nutres.2011.01.002

[131] Chapter 4: Oilseeds and Oilseed Products. Agricultural Outlook 2018-2027. OECD-FAO [Internet]. 2018. pp. 127-274. Available from: http://www.agri-outlook.org/commodities/Agricultural-Outlook-2018-Oilseeds.pdf [Accessed: 29 July 2019]

[132] Hemingway J, Eskandari M, Rajcan I. Genetic and environmental effects on fatty acid composition in soybeans with potential use in automotive industry. Crop Science. 2015;**55**:1-11. DOI: 10.2135/cropsci2014.06.0425

[133] Zhang D, Zhang H, Hu Z, Chu S, Yu K, Lv L, et al. Artificial selection on GmOLEO1 contributes to the increase in seed oil during soybean domestication. PLOS Genetics: A Peer-Reviewed Open-Access Journal. 2019;**15**:7. DOI: 10.1371/journal.pgen.1008267

[134] Panthee D, Pantalone V, West D, Saxton A, Sams C. Quantitative trait loci for seed protein and oil concentration, and seed size in soybean. Crop Science. 2005;**45**:2015-2022. DOI: 10.2135/cropsci2004.0720

[135] Mao T, Jiang Z, Han Y, Teng W, Zhao X, Li W. Identification of quantitative trait loci underlying seed protein and oil contents of soybean across multi-genetic backgrounds and environments. Plant Breeding. 2013;**132**:630-641. DOI: 10.1111/pbr.12091

[136] Lee S, Van K, Sung M, Nelson R, La Mantia J, McHale LK, et al. Genome-wide association study of seed protein, oil and amino acid contents in soybean from maturity groups I to IV. Theoretical and Applied Genetics. 2019;**132**(6):1639-1659. DOI: 10.1007/s00122-019-03304-5

[137] Ren C, Bilyeu KD, Beuselinck P. Composition, vigor, and proteome of mature soybean seeds developed under high temperature. Crop Science. 2009;**49**:1010-1022. DOI: 10.2135/cropsci2008.05.0247

[138] Mourtzinis S, Gaspar AP, Naeve SL, Conley SP. Planting date, maturity, and temperature effects on soybean seed yield and composition. Agronomy Journal. 2017;**109**:2040-2049. DOI: 10.2134/agronj2017.05.0247

[139] Kumar V, Rani A, Solanki S, Hussain SM. Influence of growing environment on the biochemical composition and physical characteristics of soybean seed. Journal of Food Composition and Analysis. 2006;**19**:188-195. DOI: 10.1016/j.jfca.2005.06.005

[140] Pipolo AE, Sinclair TR, Camara GMS. Effects of temperature on oil and protein concentration in soybean seed cultured in vitro. Annals of Applied Biology. 2004;**144**:71-76. DOI: 10.1111/j.1744-7348.2004.tb00318.x

[141] Rotundo JL, Westgate ME. Meta-analysis of environmental effects on soybean seed composition. Field Crops Research. 2009;**110**:147-156. DOI: 10.1016/j.fcr.2008.07.012

[142] Assefa Y, Purcell LC, Salmeron M, Naeve S, Casteel SN, Kovács P, et al. Assessing variation in US soybean seed composition (protein and oil). Frontiers in Plant Science. 2019;**10**:298. DOI: 10.3389/fpls.2019.00298

[143] Wood CW, Torbert HA, Weaver DB. Nitrogen fertilizer effects on soybean growth, yield, and seed composition. Journal of Production Agriculture. 1993;**6**:354-360. DOI: 10.2134/jpa1993.0354

[144] Bellaloui N, Bruns HA, Gillen AM, Abbas HK, Zablotowicz RM, Mengistu A, et al. Soybean seed protein, oil, fatty acids, and mineral composition as influenced by soybean-corn rotation. Agricultural Sciences. 2010;**1**:102-109. DOI: 10.4236/as.2010.13013

[145] Robinson AP, Conley SP, Volenec JJ, Santini JB. Analysis of high yielding, early-planted soybean in Indiana. Agronomy Journal. 2009;**101**:131-139. DOI: 10.2134/agronj2008.0014x

[146] Bellaloui N, Bruns HA, Abbas HK, Mengistu A, Fisher DK, Reddy KN. Agricultural practices altered soybean seed protein, oil, fatty acids, sugars, and minerals in the Midsouth USA. Frontiers in Plant Science. 2015;**6**:31. DOI: 10.3389/fpls.2015.00031

[147] Jaureguy LM, Rodriguez FL, Zhang L, Chen P, Brye K, Oosterhuis D, et al. Planting date and delayed harvest effects on soybean seed composition. Crop Science. 2013;**53**:2162-2175. DOI: 10.2135/cropsci2012.12.0683

[148] Fan S, Li B, Yu F, Han F, Yan S, Wang L, et al. Analysis of additive and epistatic quantitative trait loci underlying fatty acid concentrations in soybean seeds across multiple environments. Euphytica. 2015;**206**:689-700

[149] Xue HQ, Upchurch RG, Kwanyuen P. Relationships between oleic and linoleic acid content and seed colonization by *Cercospora kikuchii* and *Diaporthe phaseolorum*. Plant Disease. 2008;**92**:1038-1042. DOI: 10.1094/PDIS-92-7-1038

[150] Kumar V, Rani A, Joshi OP. Fatty acid profile released cultivars of Indian soybean with special reference to identification of high comparatively low linolenic and high oleic acid cultivars. The Indian Journal of Agricultural Sciences. 2004;**74**:388-391

[151] Fehr WR, Curtiss CF. Breeding for fatty acid composition of soybean oil. In: 7th World Soybean Research Conference, 4th International Soybean Processing and Utilization Conference, 3rd Brazilian Soybean Congress. 29-5 March; Foz Do Iguacu, Brazil; 2004. pp. 815-821

[152] Priolli RHG, Carvalho CRL, Bajay MM, Pinheiro JB, Vello NA. Genome analysis to identify SNPs associated with oil content and fatty acid components in soybean. Euphytica. 2019;**215**:54. DOI: 10.1007/s10681-019-2378-5

[153] Xia C. Study on oil content and fatty acid composition of soybean cultivars cultivated in Heliongjiang province in China. Soybean Genetics Newsletter. 1995;**22**:99-102

[154] Thapa R, Carrero-Colón M, Addo-Quaye C, Held J, Dilkes B, Hudson KA. New alleles of FAD3A lower the linolenic acid content of soybean seeds. Crop Science. 2018;**58**:1-6. DOI: 10.2135/cropsci2017.08.0490

[155] Zhang J, Wang X, Lu Y, Bhusal SJ, Song Q, Cregan PB, et al. Genome-wide scan for seed composition provides insights into soybean quality improvement and the impacts of domestication and breeding. Molecular Plant - Cell Press. 2018;**11**:460-472. DOI: 10.1016/j.molp.2017.12.016

[156] Cao Y, Li S, Wang Z, Chang F, Kong J, Gai J, et al. Identification of major quantitative trait loci for seed oil content in soybeans by combining linkage and genome-wide association mapping. Frontiers in Plant Science. 2017;**8**:1222. DOI: 10.3389/fpls.2017.01222

[157] Van K, McHale LK. Meta-analyses of QTLs associated with protein and oil contents and compositions in soybean [*Glycine max* (L.) Merr.] seed. International Journal of Molecular Sciences. 2017;**18**:pii: E1180. DOI: 10.3390/ijms18061180

[158] Li YH, Reif JC, Hong HL, Li HH, Liu ZX, Ma YS, et al. Genome-wide association mapping of QTL underlying seed oil and protein contents of a diverse panel of soybean accessions. Plant Science: An International Journal of Experimental Plant Biology. 2018;**266**:95-101. DOI: 10.1016/j.plantsci.2017.04.013

[159] Roesler K, Shen B, Bermudez E, Li CJ, Hunt J, Damude HG, et al. An improved variant of soybean type 1 diacylglycerol acyltransferase increases the oil content and decreases the soluble carbohydrate content of soybeans. Plant Physiology. 2016;**171**:878-893. DOI: 10.1104/pp.16.00315

[160] Li QT, Lu X, Song QX, Chen HW, Wei W, Tao JJ, et al. Selection for a zinc-finger protein contributes to seed oil increase during soybean domestication. Plant Physiology. 2017;**173**:2208-2224. DOI: 10.1104/pp.16.01610

[161] Hou A, Chen P, Alloatti J, Li D, Mozzoni L, Zhang B, et al. Genetic variability of seed sugar content in worldwide soybean germplasm collections. Crop Science. 2009;**49**:903-912. DOI: 10.2135/cropsci2008.05.0256

[162] Gulewicz P, Martinez-Villaluenga C, Kasprowicz-Potocka M, Frias J. Non-nutritive compounds in Fabaceae family seeds and the improvement of their nutritional quality

by traditional processing—A review. Polish Journal of Food and Nutritional Sciences. 2014;**64**:75-89. DOI: 10.2478/v10222-012-0098-9

[163] Qiu D, Vuong T, Valliyodan B, Shi H, Guo B, Shannon JG, et al. Identification and characterization of a stachyose synthase gene controlling reduced stachyose content in soybean. Theoretical and Applied Genetics. 2015;**128**:2167-2176. DOI: 10.1007/s00122-015-2575-0

[164] Jones DA, DuPont MS, Ambrose MJ, Frias J, Hedley CL. The discovery of compositional variation for the raffinose family of oligosaccharides in pea seeds. Seed Science Research. 1999;**9**:305-310. DOI: 10.1017/S0960258599000318

[165] Zheng R, Yang L, Zhou X, Zhu C, Shu X, Wu X, et al. Effect of soybean oligosaccharides on immunity and TLR2-NF-κB signal pathway response for weanling pigs. Journal of Food, Agriculture and Environment. 2012;**10**:273-279

[166] Kim HK, Kang ST, Cho JH, Choung MG, Suh DY. Quantitative trait loci associated with oligosaccharides and sucrose contents in soybean (*Glycine max* L.). Journal of Plant Biology. 2005;**48**:106-112

[167] Kim HK, Kang ST, Oh KW. Mapping of putative quantitative trait loci controlling the total oligosaccharide and sucrose content of *Glycine max* seeds. Journal of Plant Research. 2016;**119**:533-538. DOI: 10.1007/s10265-006-0004-9

[168] Mozzoni L, Chen P. Correlations of yield and quality traits between immature and mature seed stages of edamame soybean. Journal of Crop Improvement. 2019;**33**:67-82. DOI: 10.1080/15427528.2018.1542366

[169] Geater CW, Fehr WR, Wilson LA. Association of soybean seed traits with physical properties of natto. Crop Science. 2000;**40**:1529-1534. DOI: 10.2135/cropsci2000.4061529x

[170] Taira H. Quality of soybeans for processed foods in Japan. Japan Agricultural Research Quarterly. 1990;**24**:224-230

[171] Maughan PJ, Saghai Maroof MA, Buss GR. Identification and quantitative trait loci controlling sucrose content in soybean (*Glycine max*). Molecular Breeding. 2000;**6**:105-111. DOI: 10.1023/A:1009628614988

[172] Matoša Kočar M, Vila S, Petrović S, Rebekić A, Sudarić A, Josipović A, et al. Assessment of phenotypic variability of saccharides in soybean genotypes suitable for growing in Europe. Journal of Central European Agriculture. (accepted manuscript)

[173] Wolf RB, Cavins JF, Kleiman R, Black LT. Effect of temperature on soybean seed constituents: Oil, protein, moisture, fatty acids, amino acids and sugars. Journal of the American Oil Chemists' Society. 1982;**59**:230-232. DOI: 10.1007/BF02582182

[174] Openshaw SJ, Hadley HH. Maternal effects on sugar content in soybean seeds. Crop Science. 1978;**18**:581-584. DOI: 10.2135/cropsci1978.0011183X001800040014x

[175] Neus JD, Fehr WR, Schnebly SR. Agronomic and seed characteristics of soybean with reduced raffinose and stachyose. Crop Science. 2005;**45**:589-592. DOI: 10.2135/cropsci2005.0589

[176] Mozzoni L, Shi A, Chen P. Genetic analysis of high sucrose, low raffinose, and low stachyose content in v99-5089 soybean seeds. Journal of Crop Improvement. 2013;**27**:606-616. DOI: 10.1080/15427528.2013.812998

[177] Akond M, Liu S, Kantartzi SK, Meksem K, Bellaloui N, Lightfoot DA, et al. Quantitative trait loci underlying seed sugars content in "MD96-5722" by "Spencer" recombinant inbred line population of soybean. Food and Nutrition Sciences. 2015;**6**:964-973. DOI: 10.4236/fns.2015.611100

[178] Hartwig EE, Kuo TM, Kenty MM. Seed protein and its relationship to soluble sugars in soybean. Crop Science. 1997;**37**:770-773. DOI: 10.2135/cropsci1994.0011183X003400050001x

[179] Krober OA, Carter JL. Quantitative interrelationships of protein and non-protein constituents of soybeans. Crop Science. 1962;**2**:171-172

[180] Openshaw SJ, Hadley HH. Selection to modify sugar content of soybean seeds. Crop Science. 1981;**21**:805-808. DOI: 10.2135/cropsci1981.0011183X002100060002x

[181] Geater CW, Fehr WR. Association of total sugar content with other seed traits of diverse soybean cultivars. Crop Science. 2000;**40**:1552-1555. DOI: 10.2135/cropsci2000.4061552x

[182] Hollung K, Øverland M, Hrustić M, Sekulić P, Miladinović J, Martens H, et al. Evaluation of nonstarch polysaccharides and oligosaccharide content of different soybean varieties (*Glycine max*) by near-infrared spectroscopy and proteomics. Journal of Agriculture and Food Chemistry. 2005;**53**:9112-9121. DOI: 10.1021/jf051438r

[183] Yang K, Ko JM, Joung Ha T, Lee JH, Baek IY, Yang TY, et al. Development of molecular markers for low raffinose and stachyose in Korean soybean cultivars. Plant Breeding and Biotechnology. 2014;**2**:151-157. DOI: 10.9787/PBB.2014.2.2.151

[184] Bhagwat S, Haytowitz DB, Holden JM. USDA Database for the Isoflavone Content of Selected Foods, Release 2.0. Beltsville, MD, USA. Department of Agriculture, Agricultural Research Service, Nutrient Data Laboratory. DOI: 10.15482/USDA.ADC/1209670. Available from: http://www.ars.usda.gov/nutrientdata/isoflav

[185] Kim EH, Ro HM, Kim SL, Kim HS, Chung IM. Analysis of isoflavone, phenolic, soyasapogenol, and tocopherol compounds in soybean (*Glycine max* (L.) Merrill) germplasms of different seed weights and origins. Journal of Agricultural and Food Chemistry. 2012;**60**:6045-6055. DOI: 10.1021/jf300463f

[186] Jeng TL, Shih YJ, Wu MT, Sung JM. Comparisons of flavonoids and anti-oxidative activities in seed coat, embryonic axis and cotyledon of black soybeans. Food Chemistry. 2010;**123**:1112-1116. DOI: 10.1016/j.foodchem.2010.05.070

[187] Lee SJ, Seguin P, Kim JJ, Moon HI, Ro HM, Kim EH, et al. Isoflavones in Korean soybeans differing in seed coat and cotyledon color. Journal of Food Composition and Analysis. 2010;**23**:160-165. DOI: 10.1016/j.jfca.2009.08.005

[188] Li Y, Kong D, Bao B, Ahmad A, Sarkar FH. Induction of cancer cell death by isoflavone: The role of multiple signalling pathways. Nutrition. 2011;**3**:877-896. DOI: 10.3390/nu3100877

[189] Pregelj L, McLanders JR, Gresshoff PM, Schenk PM. Transcription profiling of the isoflavone phenylpropanoid pathway in soybean in response to *Bradyrhizobium japonicum* inoculation. Functional Plant Biology. 2011;**38**:13-24. DOI: 10.1071/FP10093

[190] Meng FL, Han YP, Teng WL, Li YG. Li WB. QTL underlying the resistance to soybean aphid (*Aphis glycines* Matsumura) through isoflavone-mediated antibiosis in soybean cultivar 'Zhongdou 270'. Theoretical and Applied Genetics. 2011;**123**:1459-1465. DOI: 10.1007/s00122-011-1680-y

[191] Messina MJ. Legumes and soybeans: Overview of their nutritional profiles and health effects. The American Journal of Clinical Nutrition. 1999;**70**:439-450. DOI: 10.1093/ajcn/70.3.439s

[192] Gutierrez-Gonzalez JJ, Vuong TD, Zhong R, Yu O, Lee JD, Shannon G, et al. Major locus and other novel additive and epistatic loci involved in modulation of isoflavone concentration in soybean seeds. Theoretical and Applied Genetics. 2011;**123**:1375-1385. DOI: 10.1007/s00122-011-1673-x

[193] Akond AGM, Richard B, Ragin B, Herrera H, Kaodi U, Akbay C, et al. Additional quantitative trait loci and candidate genes for seed isoflavone content in soybean. The Journal of Agricultural Science. 2013;**5**:20-33. DOI: 10.5539/jas.v5n11p2

[194] Adie MM, Krisnawati A, Harnowo D. Agronomic characteristic and nutrient content from several soybean promising lines with high isoflavones. Procedia Food Science. 2015;**3**:348-354. DOI: 10.1016/j.profoo.2015.01.038

[195] Matoša Kočar M, Vila S, Petrović S, Rebekić A, Sudarić A, Duvnjak T, et al. Isoflavone concentrations in soybeans suitable for growing in Europe. Genetika-Belgrade. 2019;**51**(1):47-59. DOI: 10.2298/GENSR1901047M

[196] Tsukamoto C, Shimada S, Igita K, Kudou S, Kokubun M, Okubo K, et al. Factors affecting isoflavone content in soybean seeds: Changes in isoflavones, saponins, and composition of fatty acids at different temperatures during seed development. Journal of Agricultural and Food Chemistry. 1995;**43**:1184-1192. DOI: 10.1021/jf00053a012

[197] Caldwell CR, Britz SJ, Mirecki RM. Effect of temperature, elevated carbon dioxide, and drought during seed development on the isoflavone content of dwarf soybean (*Glycine max* (L.) Merrill) grown in controlled environments. Journal of Agricultural and Food Chemistry. 2005;**53**:1125-1129. DOI: 10.1021/jf0355351

[198] Lozovaya VV, Lygin AV, Ulanov AV, Nelson RL, Dayde J, Widholm JM. Effect of temperature and soil moisture status during seed development on soybean seed isoflavone concentration and composition. Crop Science. 2005;**45**:1934-1940. DOI: 10.2135/cropsci2004.0567

[199] Murphy SE, Lee EA, Woodrow L, Seguin P, Kumar J, Rajcan I, et al. Association of seed and agronomic traits with isoflavone levels in soybean. Canadian Journal of Plant Science. 2009;**89**:477-484. DOI: 10.4141/CJPS08148

[200] Carrera CS, Dardanelli JL. Changes in the relationship between temperature during the seed-filling period and soya bean seed isoflavones under water-deficit conditions. Journal of Agronomy and Crop Science. 2015;**202**:421-432. DOI: 10.1111/jac.12147

[201] Morrison MJ, Cover ER, Saleem MF, McLaughlin NB, Frégeau-Reid J, Ma BL, et al. Seasonal changes in temperature and precipitation influence isoflavone concentration in short-season soybean. Field Crops Research. 2010;**117**:113-121. DOI: 10.1016/j.fcr.2010.02.005

[202] Rasolohery CA, Berger M, Lygin AV, Lozovaya VV, Nelson RL, Daydé J. Effect of temperature and water availability during late maturation of the soybean seed on germ and cotyledon isoflavone content and composition. Journal of the Science of Food and Agriculture. 2008;**88**:218-228. DOI: 10.1002/jsfa.3075

[203] Hasanah Y, Chairun Nisa T, Armidin H, Hanum H. Isoflavone content of soybean [*Glycine max* (L). Merr.] cultivars with different nitrogen sources and growing season under dry land conditions. Journal of Agriculture and Environment for International Development. 2015;**109**:5-17. DOI: 10.12895/jaeid.20151.216

[204] Wang C, Sherrard M, Pagadala S, Wixon R, Scott RA. Isoflavone content among maturity group 0 to II soybeans. Journal of the American Oil Chemists' Society. 2000;**77**:483-487. DOI: 10.1007/s11746-000-0077-6

[205] Primomo VS, Poysa V, Ablett GR, Jackson CJ, Rajcan I. Agronomic performance of recombinant inbred line populations segregating for isoflavone content in soybean seeds. Crop Science. 2005;**45**:2203-2211. DOI: 10.2135/cropsci2004.0610

[206] Hoeck JA, Fehr WR, Murphy PA, Welke GA. Influence of genotype and environment on isoflavone contents of soybean. Crop Science. 2000;**40**:48-51. DOI: 10.2135/cropsci2000.40148x

[207] Akond AGM, Liu S, Kantartzi SK, Meksem K, Bellaloui N, Lightfoot DA, et al. Quantitative trait loci for seed isoflavone contents in 'MD96-5722' by 'Spencer' recombinant inbred lines of soybean. Journal of Agricultural and Food Chemistry. 2014;**62**:1464-1468. DOI: 10.1021/jf4040173

[208] Cvejić J, Tepavčević V, Bursać M, Miladinović J, Malenčić Đ. Isoflavone composition in F1 soybean progenies. Food Research International. 2011;**44**:2698-2702. DOI: 10.1016/j.foodres.2011.05.024

[209] Seguin P, Zheng W, Smith DL, Deng W. Isoflavone content of soybean cultivars grown in eastern Canada. Journal of the Science of Food and Agriculture. 2004;**84**:1327-1332. DOI: 10.1002/jsfa.1825

[210] Yin X, Vyn TJ. Relationships of isoflavone, oil, and protein in seed with yield of soybean. Agronomy Journal. 2005;**97**:1314-1321. DOI: 10.2134/agronj2004.0316

[211] Al-Tawaha AM, Seguin P. Seeding date, row spacing, and weed effects on soybean isoflavone concentrations and other seed characteristics. Canadian Journal of Plant Science. 2006;**86**:1079-1087. DOI: 10.4141/P06-043

[212] Charron CS, Allen FL, Johnson RD, Pantalone VR, Sams CE. Correlations of oil and protein with isoflavone concentration in soybean (*Glycine max* (L.) Merr.). Journal of Agricultural and Food Chemistry. 2005;**53**:7128-7135. DOI: 10.1021/jf050610o

[213] Wang HJ, Murphy PA. Isoflavone composition of American and Japanese soybeans in Iowa: Effects of variety, crop year, and location. Journal of Agricultural and Food Chemistry. 1994;**42**:1674-1677. DOI: 10.1021/jf00044a017

[214] Bursać M, Atanacković Krstonošić M, Miladinović J, Malenčić Đ, Gvozdenović LJ, Hogervorst J, et al. Isoflavone composition, total phenolic content and antioxidant capacity of soybeans with colored seed coat. Natural Product Communications. 2017;**12**:475-640

[215] Sumardi D, Pancoro A, Yulia E, Musfiroh I, Prasetiyono J, Karuniawan A, et al. Potential of local black soybean as a source of the isoflavones daidzein and genistein. International Food Research Journal. 2017;**24**:2140-2145

[216] Tepavčević V, Atanacković M, Miladinović J, Malencić D, Popović J, Cvejić J. Isoflavone composition, total polyphenolic content, and antioxidant activity in soybeans of different origin. Journal of Medicinal Food. 2010;**13**:657-664. DOI: 10.1089/jmf.2009.0050

[217] Morito K, Hirose T, Kinjo J, Hirakawa T, Okawa M, Nohara T, et al. Interaction of phytoestrogens with estrogen receptors alpha and beta. Biological & Pharmaceutical Bulletin. 2001;**24**:351-356. DOI: 10.1248/bpb.24.351

[218] Wang Y, Han Y, Teng W, Zhao X, Li Y, Wu L, et al. Expression quantitative trait loci infer the regulation of isoflavone accumulation in soybean (*Glycine max* L. Merr.) seed. BMC Genomics. 2014;**15**:680. DOI: 10.1186/1471-2164-15-680

[219] Wang Y, Han Y, Zhao X, Li Y, Teng W, Li D, et al. Mapping isoflavone QTL with main, epistatic and QTL × environment effects in recombinant inbred lines of soybean. PLoS One. 2015;**10**(3):e0118447. DOI: 10.1371/journal.pone.0118447

Chapter 4

Respiratory Health Risks from Exposure to Dust from Soybean and Its Products

Howard J. Mason

Additional information is available at the end of the chapter

http://dx.doi.org/10.5772/intechopen.92839

Abstract

Soybean and its processed forms have become an increasingly important part of agriculture, where they are widely used as an animal feedstuff and in an extensive range of human food products. This entails transportation from producer countries, largely the USA and South America, to importer destinations such as the EU and China. Soya is recognised as a dietary allergen, containing a number of identified allergenic proteins. Inhaled soya dust generated by occupational activities also causes respiratory health problems. Reports of "asthma epidemics" in harbour cities identified ill health in the community but were related to occupational activities of unloading/loading bulk soya without appropriate dust control measures. Inhaled allergens in already-sensitised individuals can cause a range of allergic symptoms in the eyes, nose, and respiratory tract, including occupational asthma (OA). Soybean dust can also cause health problems related to lung irritancy. Endotoxin and fungal contamination associated with soya are also potential respiratory hazards. This chapter reviews published data on ill health and symptoms from airborne exposure to allergens in dust from soybean and derivative products and the levels of exposure to dust and major allergens from a range of occupational activities. Other potential health hazards associated with soybean, such as endotoxin and fungal contamination, are also highlighted.

Keywords: soy, soybean, soya, asthmagen, allergy, occupational asthma, endotoxin

1. Introduction

Soybean (*Glycine max* (L.) Merr.) has become one of the most important, versatile globally traded commodities, being a widely used source of protein, oil, and biofuel. Its uses include as a source of protein and fibre for livestock and an alternative to meat and dairy products in humans. Soya products are also increasingly used widely in the food industry, in particular as texturisers, emulsifiers, and protein fillers; soya flour is often added to bakery products, such as bread, biscuits, pastry, etc. Soybean oil is the second largest source of vegetable oil globally and is also used in products such as biodiesel and detergents.

Soybeans are crushed to form meal, typically used in animal feed, and oil. The hull or husk of the soybean is a by-product of soybean oil and meal production where the beans are de-hulled prior to crushing. Soya hull is also internationally traded as an animal feedstuff, providing a good source of digestible fibre, albeit of lower protein content of soya meal.

The EU imported about 18 million metric tons of soya in 2018 [1]. Approximately 90% of these imports are used to feed livestock and reflect about 28% of global soya imports. China imports approximately 88 million metric tons. Although the USA remains the largest exporter of soya, projected export growth is concentrated in South America, particularly Brazil, Argentina, Paraguay, and Bolivia. The UK imports some 3 million tons annually with more than 70% directly from Argentina and Brazil. There is also an inter-trade within Europe, with the Netherlands being an important hub. The UK imports approximately two thirds as soya meal/hulls and one third as soybeans [2]. The UK only imports a relatively small quantity of soya oil, approximately 200,000 tons. The UK does not produce biofuels to any extent from imported soya.

Therefore, there is considerable bulk transportation by sea, involving handling at ports equipped to handle bulk grains and foodstuffs. Thereafter there is onward transportation for use in the animal feed industry, further processing, and the human food sector.

However, soya is not without associated risks to health. Soya products are recognised as one of the EU's 14 major food allergens and listed in Annex II of the EU Regulation 1169/2011 on labelling of foods and UK equivalent domestic legislation [3]. It is also listed as a major food allergen by the FDA (USA) labelling regulations. As soybean and its products are used in many processed foods, it is difficult for the allergic consumer to avoid and is often classified as a "hidden allergen". Additionally, evidence from a number of sources identify proteins found in soybean and its products as respiratory allergens capable of producing a range of ocular and upper and lower respiratory symptoms, including asthma.

This chapter focuses on both published evidence and our own studies related to the respiratory risk from airborne dusts related to soya.

2. Respiratory risks from soybeans

2.1. Soybean and its products

Soybean (*Glycine max* (L.) Merr.) or soya bean is the edible seed of an annual legume of the pea family (Fabaceae). The hull or husk of the mature bean is hard and water-resistant and protects the cotyledon of the seed from damage.

The major forms of soya usually encountered in end-user countries in the EU are:

1. Soybean, after removal of hull covering the bean, containing about 40% protein and 20% fat/lipid.

2. Soya meal (see **Figure 1**). This may be of two forms: pure meal produced after de-hulling and possibly extraction of oil or with subsequent added hull to extend the product. Soybean meal made from de-hulled beans has a total protein content of approximately 40–49% and 3% fibre.

3. Soybean hull, these are often pelletised as a commercial product to make a more handle-able, less dusty product (see **Figure 1**). The protein content of hull is around 9–19%, with a fibre content of 53–74%. The proteins in hull tend to be of lower molecular weight than those in pure soya meal (**Figure 2**).

4. Soya oil is produced by crushing and/or chemical extraction. Soya oil, particularly the more highly purified, is considered less allergenic due to the low concentration of soya proteins within it [4, 5]. It is used widely in food processing.

5. Soya flour—milled in a similar way to cereal flour (e.g. wheat, rye). Flours from various cereals have the propensity to be "dusty", and the control of their handling is necessary to prevent airborne exposure to flour dust and consequent health effects [6, 7]. Soya flour has become increasingly used in food processing. Allergens in soya flour have been identified and characterised [8].

Figure 2 shows electrophoresis gels of extracts of a soya hull and soya meal, respectively, after extraction at 10% w/v using 0.1% Tween 20 in phosphate buffered saline. These gels separate proteins on the basis of their molecular weights. The patterns of proteins in soya hull show considerable differences to soya meal. There is a predominance of high molecular weight proteins in meal in comparison with hull where the majority of proteins appear to be less

Figure 1. The left hand image shows an image of a soya meal imported in the UK. The right-hand image shows a sample of soya hull imported into the UK. The pelletised hull material shows some evidence of breakdown, probably due to compaction in the hold of the ship.

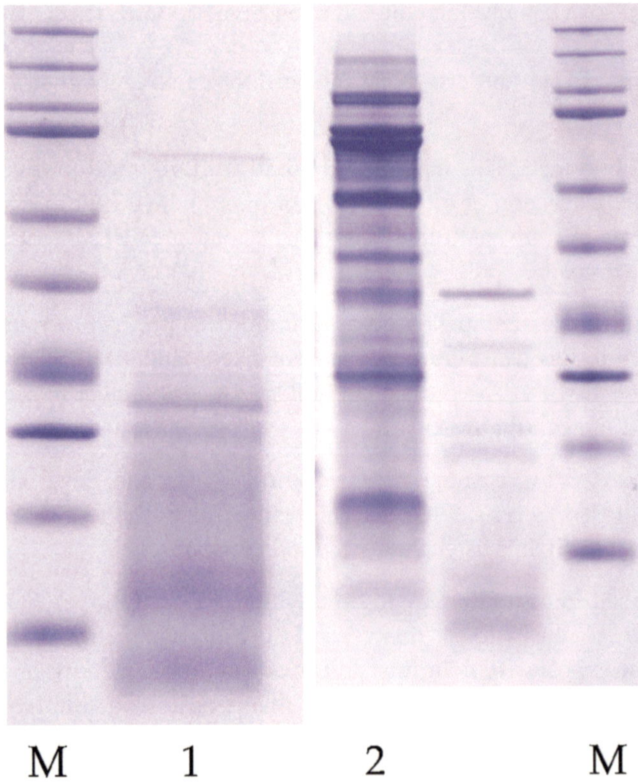

Figure 2. Left-hand gel shows a Coomassie blue-stained reducing gel of an extracted soya hull material, lane labelled 1. Right-hand gel shows a similarly stained gel of an extract of soya meal, lane labelled 2. Molecular weight marker bands for lanes marked as M have molecular weights from the top of 200, 150, 100, 75, 50, 37, 25, 20, 15, and 10 kDa.

than 23 kDa. For some soya meal products, hull is reintroduced to adjust the overall protein content, so differing soya meal imports may contain differing levels of hull proteins.

However not all proteins are allergenic, in terms of sensitising an individual's immune system to provoke an exaggerated IgE-mediated immune response on subsequent exposure to the same protein, i.e. a type 1 allergic response. Allergenic proteins appear to be restricted to classes or families of proteins based on their structural and functional properties [9–12].

2.2. Overview of intrinsic and extrinsic respiratory risk factors in soybean

However, besides intrinsic, specific allergenic proteins, covered in the next section, there are other "contaminants" or extrinsic material that may be associated with soya products and possibly lead to respiratory illnesses or symptoms, if inhaled. These include:

- Endotoxin is a pyrogenic lipopolysaccharide and a component of the exterior cell wall of gram-negative bacteria, like *E. coli*. High concentrations of airborne endotoxin can cause respiratory inflammation, symptoms, and lung function decline [13–16]. The Netherlands has set a suggested health-based exposure limit for airborne endotoxin [17]. Endotoxin has been found to be extractable from soya meal and husk and becomes airborne when handling bulk [18–21].

- β-Glucans are naturally occurring polysaccharides, being constituents of the cell wall of certain pathogenic bacteria and almost all fungi. Their measurement in airborne samples has been used as an indicator of total fungal exposure. β-Glucans have been linked to activating macrophages, neutrophils, monocytes, and NK cells, thus involving the innate and adaptive immune systems. Biological activity seems related to their degree of branching and molecular weight; greater branching gives rise to greater biological activity, with the $(1 \rightarrow 3)$ chain essential in the induction of immune responses [22].
- Fungi such as *Aspergillus* spp., particularly the *A. glaucus* group, and *Penicillium* spp. are known as storage moulds. Contamination of batches of soya with uncontrolled fungal growth, particularly *Aspergillus* spp. and *Penicillium* spp., leads to spoilage. Certain toxicogenic *Aspergillus* species under the right conditions of moisture and temperature can lead to the production of mycotoxins and carcinogenic aflatoxins [23, 24]. In addition, *Aspergillus* and *Penicillium* species are allergenic and can also cause hypersensitivity pneumonitis (HP), also known as extrinsic allergic alveolitis (EAA). *Aspergillus fumigatus* can produce significant numbers of conidia (spores) containing allergenic proteins, e.g. Asp f 1, and in immune-compromised humans is the most common life-threatening, opportunistic fungal pathogen. Nonetheless, several strains of *Aspergillus* are used in the controlled fermentation of soya to produce soy sauce, including *A. oryzae*. Alpha amylase from this fungal source is used as an additive improver in cereal flour and associated with significant sensitisation in bakers [25, 26].
- Organic dust with no identifiable toxic properties can cause irritation and inflammatory responses in the lungs if the particles are small enough. Larger dust particles will lodge in the nasal passage or the throat and be cleared from the body. Particles of less than 10 μm aerodynamic diameter can enter the lungs past the bronchus, and particles less than 4 μm can reach the alveoli deep in the lungs, producing significant lower respiratory tract symptoms. Limited evidence suggests that this mechanism may be relevant for soya dust [27]. Organic dust toxic syndrome (ODTS) and EAA are distinct pathological entities associated with smaller particles below 5 μm [28]. Asthmatic reactions are generally provoked by particle sizes of 5–10 μm [29].

Of these intrinsic and extrinsic factors potentially associated with soya, it is the health effects from exposure to intrinsic soya allergens that are underpinned by significant scientific evidence. This will be the major thrust of the remainder of this chapter.

2.3. Allergens in soybean

A number of allergens have been identified and characterised in soya and its products. A 2012 OECD document on soybean allergens lists 15 proteins designated as allergens, largely derived from one literature review [12]. However there has been a criticism about the lack of evidence for some of these "putative" allergens [30]. A non-exhaustive list of allergens is shown in the **Table 1** below. Many of the allergens were identified from a food perspective, with subsequent work to produce "hypoallergenic" cultivars [12]. While there has been considerable research on genetically modified soya with lower levels of endogenous major food allergens, a large natural variation (9–15-fold) in the levels of Gly m 4, Gly m 5, Gly m 6, Gly m Bd 28 k, and Gly m Bd 30 k has also been identified [31].

However, a much smaller number of the allergens in **Table 1** have been implicated in terms of airborne exposure during occupational practices and associated health effects.

Allergen	Description	Comments
Gly m 1	Hydrophobic soybean protein. MW 7–8 kDa with two isoforms	Abundant in soybean dust. Husk and pods are a rich source. Implicated in epidemic asthma outbreaks in harbour cities caused by soy dust [32, 33]
Gly m 2	MW 8 kDa protein with a pI 6. A member of the defensin family	Gly m 2 is abundant in soya husk and implicated in epidemic asthma outbreaks in Spanish dock cities [34]. Shows some homology with a storage protein in the cotyledon of cowpea and green pea
Gly m 3	MW 12–15 kDa protein	A profilin type of allergen. Shows some cross-reactivity with birch profilin [11, 35]
Gly m 4	MW 17 kDa. Homolog of Bet v 1, a birch allergen	Implicated as the major allergen where patients are allergic to birch pollen and have soy allergy [36]
Gly m 8	MW 28 kDa. 2S albumin	2S albumins [37]. Some homology with Ara h 2, a peanut allergen. Identified as a food allergen
Gly m 39kD	MW 39 kDa	P39 protein was detectable only in the fully mature dry seed distributed in the matrix of the protein storage vacuoles [38]
Gly m Bd28K	MW 28 kDa. A vicilin-like glycoprotein	A major food allergen [39]
Gly m Bd30K	MW 30–34 kDa protein, a thiol protease of the papain superfamily	A soybean oil body-associated glycoprotein, shows 30% sequence homology with Der p 1, a major allergen of house dust mite. An important dietary allergen, widely known as P34 [40]
Gly m Bd 60 K	MW 63–67 kDa protein	An alpha subunit of beta-conglycinin well-known as a major soybean storage protein. Major food allergen
Gly m TI	MW 20 kDa, a trypsin inhibitor	Has been implicated as a workplace inhalant allergen in bakers [41]. Found in the seed and soya flour
Gly m 5	β-Conglycinin, three isoallergens	Seed storage protein. Sensitisation to Gly m 5 is potentially indicative for severe allergic reactions to soy [42]
Gly m 6	Glycinin, five isoallergens	Sensitisation to Gly m 6 is potentially indicative for severe allergic reactions to soy [42]
Gly m 7	MW 76 kDa Seed biotinylated protein (SBP)	SBP may represent a class of biologically active legume allergens with structural resilience to many food-manufacturing processes [43]

MW refers to molecular weight.

Table 1. A non-exhaustive list of allergens identified in soya.

2.4. Published studies on health outcomes and airborne exposure to soya dust

Further airborne exposure to an allergen in an individual already sensitised can cause a range of symptoms affecting the eyes, nose, and upper and lower respiratory systems, including the development of occupational asthma (OA). OA is a disease characterised by variable airflow limitation and airway hyperresponsiveness due to a particular occupational environment. Two main types of OA are identified [44]. Immunological OA develops after a latent period of exposure during which the worker acquires sensitisation to the causal agent, typically involving IgE-mediated immunological sensitisation to allergenic proteins. Non-immunologic OA is usually due to irritant mechanisms associated with the cumulative effects of exposure to a workplace dust or chemicals. Both forms of OA can be serious enough to prevent an individual's continued employment in that workplace and even cause permanent disability.

The first study describing soya allergy related to dust from a soybean mill was published in 1934 [45]. In 1977 a study was published of immediate and late-onset OA in a previously non-allergic subject exposed to soya flour in the manufacture of food supplements [46]. Exposure to soya dust and soya flour has been implicated in causing OA or other respiratory health symptoms in persons working in a variety of occupations, such as farmers, millers, soybean processors, and bakers [8, 26, 27, 41, 45, 47, 48].

2.4.1. Asthma epidemics in harbour cities related to soya dust

In the 1990s, a number of scientific papers were published that investigated "asthma epidemics" in harbour cities, the cases of asthma being found in the general population. Investigation discovered these asthma cases were related to the loading or unloading of soya products. Reports were related to New Orleans, the USA [49], Cartagena, Spain [50, 51], Tarragon, Spain [52], Saint-Nazaire, France [53], Naples, Italy [54], Valencia, Coruna, Spain [55], and Barcelona, Spain [56].

The original outbreaks of asthma epidemics occurred in New Orleans, starting in 1953 and continuing for almost 20 years. Sometimes more than 200 people sought treatment in a single day at a hospital serving a largely black, poor population [57]. Initial investigations associated the outbreaks with low wind speeds but from a specific direction and together with particular climatic conditions. However, it was only in 1997, and after the investigations concerning Barcelona, that these community asthma outbreaks were specifically linked to the loading of soya (but not wheat or corn) into ships using an elevator system [49], suggesting that soy dust may be particularly asthmagenic compared with some grain dusts.

The asthma epidemics that occurred in Barcelona have been the best documented, and a considerable amount of research was expended in linking soya unloading at the docks with the asthma epidemics in the city, rather than other possible precipitating factors, such as traffic pollution, moulds, etc. [24, 34, 58–60]. From 1981 to 1987, 26 outbreaks of asthma occurred in the city of Barcelona, affecting a total of 687 subjects and causing 958 emergency room admissions and 20 deaths. Further outbreaks occurred in 1994 and 1996. The initial asthma events coincided with the unloading of soya into silos without a filter, climatic conditions of high-pressure areas, and the wind direction from the harbour to the city [61].

While it might be that very specific geo-climatic conditions were the drivers for the Barcelona and other asthma epidemics, a number of important factors emerged of wider significance. Some of which were confirmed from other studies of asthma epidemics, and some of which suggested the need for further work as follows:

- The latency period from initial unloading of soya in Barcelona to asthma outbreaks appears consistent with that of occupational asthma. Children were rarely affected in these asthma epidemics, and age appeared a risk factor [62].
- The primacy of implementing exposure control measures on the occupational processes to control dust emissions and prevent further asthma outbreaks [49, 63].
- While climatic conditions may have been important, these phenomena suggest that some soya dusts generated are of a small aerodynamic diameter with high buoyancy to travel relatively large distances and penetrate deep in the lungs.
- The allergenic material identified in Barcelona implicated glycoproteins with molecular weights lower than 14 kDa, with the major allergen identified as Gly m 1 [33, 64], localised in soybean hulls/husks (see **Table 1**). Gly m 2 was also implicated [34]. Ninety-two percent of patients in the Tarragona epidemics were sensitised to soybean hull extracts [52].
- In response to the Barcelona episodes, significant effort was put into developing immunoassays capable of quantifying the putative allergen(s) with the necessary sensitivity to measure airborne levels. As with many other aeroallergen immunoassays, they progressed from initial competitive immunoassays utilising pools of serum from sensitised individuals [65] to non-competitive, sandwich assays based on polyclonal [66] or monoclonal antibodies [67]. As found for other aeroallergens, the inhibition assays are less sensitive and give considerably higher results when compared with non-competitive sandwich immunoassays [66].
- Airborne Gly m 1 levels were measured by monoclonal sandwich immunoassay, at progressive distances from Ancona's (Italy) port, where soya is unloaded [68]. Allergen concentrations were less than 171 ng m^{-3}, whereas HSP levels (highly homologous with Gly m 1 [66]) measured by sandwich immunoassay during dockside activities in Barcelona and the UK were considerably higher [19]. Decreases in allergen away from the unloading area in Ancona were detected. Airborne Gly m 1 was not coupled with the presence of soya-carrying ships in the port, but significant relationships between allergen and meteorological parameters were found, suggesting that Gly m 1 appeared part of Ancona's atmospheric dust. The authors suggest these allergen levels seem consistent with the absence of asthma epidemic outbreaks in Ancona.
- There is evidence of genetic factors, atopy, and smoking status modifying the response to exposure to soybean dust [62, 69]. Atopy and smoking have been identified as risk factors for sensitisation and work-related respiratory symptoms with a number of other occupational allergens, e.g. bakers [70], laboratory animals workers [71], and seafood processors [72, 73].

- There is evidence of co-exposure and sensitisation to some fungi and moulds, but it does not appear to have been causative of the symptoms/illnesses. Specific IgEs in a small group of asthma epidemic (AE) patients were compared with asthmatic non-epidemic patients and non-allergic controls [24]. The AE group showed low levels of specific IgE to *A. flavus*, *A. fumigatus*, *A. glaucus*, *Penicillium notatum*, and *P. chrysogenum* but significantly lower than IgE levels against soybean hull. All the AE group were sensitised to soya hull but between 8 and 92% against the moulds (*A. flavus*, *A. nidulans*, *A. glaucus*, and *P. notatum* being predominant).

2.4.2. *Bakers and soybean flour*

Alvarez [74] showed in a small-scale study that 25% of bakers were sensitised to soybean. A review of cross-sectional studies employing skin prick tests in bakers showed that 5–77% were sensitised to soybean flour [75]. A relatively recent UK study suggested a prevalence of 21% sensitisation using similar methodology [76]. Baur [77] found 21% serological sensitisation to soybean flour in 140 bakers who had a history of greater than 6 months of employment and work-related asthma, rhinitis, and/or conjunctivitis. Two workers were shown to be sensitised to soybean lecithin, although the lecithin was possibly contaminated with low levels of soya proteins [78]. Baur [41] studied a relatively small group of bakers both sensitised to soybean and suffering workplace symptoms. Twelve were also sensitised to wheat, ten to rye, and five to alpha amylase from *A. oryzae* (FAA). The latter being an enzyme often added to flour in small quantities, but it is now regarded as a potent allergen [25]. Baur identified soya trypsin inhibitor (STI) or Gly m T1 as a major allergen, being recognised by IgE antibodies in the sera of 86% of the examined sensitised bakers. This research was one of the drivers for the Health and Safety Executive (HSE) to develop an immunoassay sensitive enough to detect airborne levels of STI from the use of soya flour and possibly other soya products [18, 19].

In a laboratory study of components of flour improvers, a representative soya flour was neither more inherently "dusty" nor showed a shift to smaller particle sizes than three different wheat flours [79]. However, although the improvers contained a higher percentage of wheat flour than soya flour, there was roughly 10-fold more extractable STI in comparison to wheat alpha amylase inhibitor (WAAI) per unit weight of improver. WAAI is a major allergen and sensitiser in bakers, with a subunit size of around 14–16 kDa and is restricted to the seed storage tissue (endosperm) [80, 81].

Quirce [82] examined four bakers or confectioners who were sensitised to both soya and wheat using skin prick tests. A positive response to STI and FAA was noted in 2/4 cases. IgE-binding bands against soya flour showed bands at molecular weights between 25 and 55 kDa and also high molecular weight IgE-binding bands against hull extract. A case study [8] of a sensitised individual presenting with asthma after 6 years of using soya flour in food processing (not a bakery) showed immunoreactivity against nine soya proteins in the molecular weight range of 15–55 kDa. Interestingly, cross-reactivity studies with other legumes demonstrated apparent immunologic identity between a component in green pea extract and a soybean protein with a molecular weight of 17 kDa [8].

Overall these data confirm that the allergens caused by soya flour are predominantly higher molecular weight proteins, whereas the asthma epidemics in harbour cities were caused by low molecular weight proteins, specifically the allergenic proteins, Gly m1 and Gly m 2.

2.4.3. Workplaces processing soya

Early investigations in Yugoslavian soya processors by Zuskin [27, 83] studied dust inhalation and respiratory symptoms after the oil had been extracted. Exposed workers showed a considerable increase in respiratory symptoms over controls, e.g. cough, nasal symptoms, and wheezing being reported by 56, 41, and 30% of workers, respectively. Most workers were smokers, and inhalable dust levels were considerable, with a mean (range) of 29.5 (7.7–59.9) mg m^{-3}. Decreases in lung function were noted over the working week and pre-shift Monday testing suggesting evidence of chronic impairment [83]. Sixteen percent showed serological evidence of specific anti soya IgE, although 68% were positive against house dust mite. Zuskin appeared to be suggesting an irritant rather than immunologic mechanism for the airways disease.

Two related studies [21, 26] investigated sensitisation, symptoms, and exposure measurements in three South African soya processing plants. These plants were producing soya flour, based on de-hulling, cooking, and finally milling. Median (range) of inhalable dust levels were 2.58 (0.24–35.02) mg m^{-3}; STI allergen levels gave a median (range) of 70 (50–2580) ng m^{-3} and were higher in the later parts of the process. There was no significant correlation between dust levels and allergen levels. Thirty-one percent of workers were current smokers, much lower than found in Zuskin's study. There were significant associations between worked-related chest tightness, nasal symptoms, and cough/chest tightness after handling soya and sensitisation to soybean. Thirty-three percent of the workers were atopic, and 14% were sensitised to soybean not containing hull allergens. Atopy but not smoking was associated with sensitisation to soybeans, confirming the association between atopy and sensitisation to occupational allergens (Section 2.4.1).

Interestingly, Harris-Roberts reported that those transferring soybeans from farms into the processing plants' silos, where soya hull would be present, had an excess of "flu-like "symptoms of fever, aching, and tiredness [26]. Such work-related, flu-like symptoms unrelated to soya sIgE levels were also noted by Cummings, but in processors not exposed to hull [84]. The biological reason for the "flu-like" symptoms is unclear. Harris-Roberts [26] hypothesised that these symptoms may suggest organic dust toxic syndrome (ODTS) in which inhaled endotoxin has been implicated [85]. Higher levels of endotoxin are found in hull rather than soybean or soya meal [18, 19, 26], but unfortunately airborne endotoxin levels were not measured in the Harris-Roberts study. Hypersensitivity pneumonitis (HP) also called extrinsic allergic alveolitis (EAA) has been reported in a single case while handling soybean as an animal feed [86]. Both ODTS and HP can give rise to similar "flu-like" symptoms some 4–12 hours after exposure. Whatever the cause or pathology, it raises the possibility of other health problems in soya-exposed workers besides those caused by IgE-mediated sensitisation.

A study was carried out in 2007 at a US soya processing plant receiving de-oiled, de-hulled, and crushed soya flakes for further processing. Concerns had been raised about asthma and other respiratory symptoms [48, 87]. Serum IgE immunoblotting studies showed multiple soya antigens, with 48, 54, and 62 kDa being most prominent, including storage proteins Gly m 5 and Gly m 6. As possibly expected, no sIgE to Gly m 1 or Gly m 2 was detected in this de-hulled material. The prevalence of soya specific IgE was 21% (versus 4% in controls), albeit only 7% gave a positive skin prick test for soya. Ten percent showed specific IgE towards storage mites. Those participants with soya-specific IgE had a threefold risk of current asthma or asthma-like symptoms and a six fold risk of work-related asthma symptoms. Thus asthmas and symptoms of asthma were associated with immunogenic nature of this de-hulled soya material. Work-related sinusitis, nasal allergies, and rash were also associated with reported mould exposure.

A single case study was reported from an animal feed factory, where for 5 years a man had been separating the soybean from hull before grinding [88]. He was atopic, although negative to storage mite. He showed a strong bronchial response to a challenge by soya hull but negative to soya flour. Unfortunately this short report is not clearer on the specific tasks being undertaken.

Heederik [89] studied sensitisation and respiratory effects in atopics and asthmatics (cases) living close to a Dutch soya oil producing factory. Soybeans and the oil product were transported by ship. Soya waste, after oil extraction, was removed by truck and noted to be "very dusty". Soybean unloading was carried out without any emission controls and caused visible dust clouds. Loading trucks with waste also caused dust clouds around the factory area, with spillages in transit. Only 11% of the cases were sensitised to soya by skin prick test, the same as in matched controls. Soya-sensitised individuals living in proximity to the factory reported more respiratory symptoms, used bronchodilators more often, and had poorer lung function after having been downwind of the factory. Airborne soya allergen, measured by competitive immunoassay, was found more frequently surrounding the factory with levels higher than in the control area but much lower than found on the factory premises. Periodic, high endotoxin concentrations close to the factory exceeded the suggested Dutch threshold level of 90 EU m^{-3} [17]. Interestingly only 14% of workers, although more highly exposed than the cases, were sensitised to soya, with 31% being atopic.

2.4.4. *Population studies in producer countries*

A study in Argentina, which is an important producer of soybeans and its products, looked at 365 cases of asthma or allergic rhinitis and 50 healthy controls. Both groups were classified as to whether they had occupational exposure to soya, were in proximity to soybean fields or grain elevators, or lived in an urban environment without obvious exposure to soybean dust [90] . The overall prevalences of sensitisation by skin prick test to soya hull in cases and controls were 15 and 0%, respectively. In the cases subdivided by exposure classification, these sensitisation prevalences were 39% (occupationally), 20% (proximity), and 8% (urban). Positive skin prick tests were higher for mites (mainly storage mites), pollen, and moulds in those positive to soya hull extract. Serological sensitisation (sIgE) to soya hull was 39 and 10% in cases and controls, respectively. The data suggest that atopic status and inhalation of soybean dust are necessary for sensitisation to soya hull. The authors opine that sensitisation to moulds could be related to contaminated soya and noted that no near-fatal or fatal asthma had occurred, unlike the situation in epidemic asthma outbreaks involving sudden exposures to soya dust. The authors suggested that their data indicates that an immunologic mechanism rather than irritancy is responsible for soybean-induced asthma in those repeatedly exposed.

2.4.5. Studies undertaken at the Science and Research Centre of the Health and Safety Executive

Three studies were undertaken by the HSE during 2012–2017. Two studies involved occupational hygiene monitoring at different UK ports handling soya. The third study was laboratory-based, investigating inherent "dustiness" in seven imported bulk soya products. Two established allergen assays were employed: a polyclonal sandwich assay for hydrophobic seed protein (HSP) established by our collaborators in Barcelona—HSP is highly homologous with the two Gly m 1 isoallergens [66, 91]—and soya trypsin inhibitor (STI) that has been implicated as a major allergen in bakers handling soya flour [41]. Endotoxin measurements were employed to establish the extent of endotoxin contamination of soya products and the levels of airborne endotoxin that workers may inhale.

One of these studies was in response to a complaint of respiratory symptoms in a workplace situated some 300 m from a dock in the South of England. This dock is used for the unloading of soya from bulk cargo ships, its storage, and onward transport to end-users [19, 92]. Essentially, this was an occupational hygiene study but also measured the levels of soya allergen at the perimeter of the dockside operation and slightly beyond. The dock is situated to the west of a city centre of some 250,000 individuals. Containers ships are emptied by dockside grab cranes into hoppers for loading of either heavy goods vehicles for onward transportation or a conveyor belt whereby the soya was transferred to storage warehouses on-site. Concerns had been raised by the stevedores and harbour managers about the unloading of a particularly dusty batch of finely ground soya meal. But generally soya dust was visually noticeable during any unloading activity of soya meal or hull.

Samples of four different soya bulks unloaded during the study were collected. One bulk had evidence of areas of gross fungal contamination, which was identified as *Aspergillus glaucus* with moderate amounts of *Aspergillus fumigatus*. Both these fungi are common on vegetation and stored agricultural material and with sufficient available water can allow for potential heavy growth. Inhalation of these fungi is also implicated as causing hypersensitivity pneumonitis. Hull was unloaded on day 1 and meal on days 2 and 3. The hull product was a pelletised material, showing evidence of breakdown (see **Figure 1**).

The hull sample had considerably more endotoxin than meal samples (**Table 2**). While the hull sample had 15-fold more HSP than the meal, the difference in STI levels between the hull and meal was much lower. **Table 3** shows the results from static air monitoring at or outside the perimeter of the dock operation. Amounts of allergen were measurable at these peripheral sampling sites. On the day of hull unloading, significant levels of the allergen associated with asthma epidemics were measured some 150–200 m in the prevailing wind direction from where soya dust clouds were being generated.

Bulk material	Endotoxin (EU g^{-1})	HSP (µg g^{-1})	STI (µg g^{-1})	Asp f 1 (µg g^{-1})	Der p 2 (µg g^{-1})
Pelletised hull	80,364	2824	798	Trace	Trace
Meal	4630	196	270	ND	ND
Meal, GM-free	1309	178	233	ND	ND

ND is non-detected.

Table 2. Amounts of allergens and endotoxin extracted from the three bulk samples not showing evidence of gross fungal contamination.

Site	Position	STI (ng m^{-3})	HSP (ng m^{-3})
Cruise ship customers' car park	170 m from nearest source (conveyor or hopper). In prevailing wind direction	13 (ND-40)	26 (11–125)
East end of building off-site	150–200 m from open sources. In prevailing wind direction	ND (ND-4)	87 (30–1300)[a]
Road entrance	100–150 m from sources of conveyor or hopper	19 (2–24)	54 (26–85)
Steps at boundary wall	150 m from hopper. In prevailing wind direction	56 (7–80)	339 (27–898)[a]

[a]High value associated with day of hull unloading.

Table 3. Allergen levels at the perimeter of an UK dock operation and, beyond, unloading soya.

Visual dust clouds outside were noted during various activities: (a) loading of hoppers; (b) loading of lorries from the hopper, see **Figure 3**; (c) the moving conveyor, which was subject to spillages; and (d) from craneage of soya out of the ship's holds, see **Figure 4**. Visual clouds of dust were also produced within the storage facility as the unloaded soya was formed into piles by a pusher loader or loaded into vehicles for onward transport. Respiratory protective equipment (FFP2 respirators) was worn by workers in the ship's hold, but was not uniformly worn elsewhere.

The median (range) of personal atmospheric monitoring sampling in workers over the 3 days of study were 130 (33–3071) ng m^{-3} and 583 (170–12,629) ng m^{-3} for STI and HSP, respectively. High allergen values were found when moving soya within the enclosed storage warehouses and within the ship's holds. Inhalable dust exposures (personal samples), expressed as 8 hour TWA, ranged from 1.2 to 4.5 mg m^{-3}; the current UK workplace exposure limit (WEL) for flour dust and grain dust is 10 mg m^{-3}. Interestingly on the day that hull was unloaded, high levels of HSP (2925 ng m^{-3}) were sampled within the crane's cab although some 50 m above the dockside.

The second HSE study addressed issues concerning the likely differences in dustiness of various bulk soya products and the categorisation of the particle size of dusts generated [18]. A rotating drum testing method has been established that can investigate the generated levels of a dust under standardised conditions that are associated with the defined inhalable, thoracic, and respirable particle size fractions [93–95]. Inhalable particles of an aerodynamic diameter (AD) ≤ 100 µm can enter the respiratory tract via the nose and mouth. Thoracic sized particles

Figure 3. The visual dust cloud from a lorry loading soya from the hopper and the crane grab depositing soya into the hopper.

Figure 4. The crane is being used to move soya from the ship's hold to the dockside hoppers. Spillages and dust clouds from the crane's grab are noticeable.

(AD < 30 μm) are defined as those small enough to penetrate past the larynx as far as the trachea and bronchial areas of the lung. Respirable particles (AD < 10 μm) can enter the deeper part of the lungs.

Essentially, a fixed amount of bulk material is rotated at a set speed and time-period in a drum with vanes that lift and drop the bulk material during rotation. A constant airflow through the drum entrains any airborne dust that is collected on an in-line series of two metal foams with different pore densities and finally a glass microfibre filter. Three replicate runs with gravimetric analysis and extraction of allergens of the foams and filter are used to calculate an average dustiness in the inhalable, thoracic, and respirable sized fractions. This technique was used to compare the intrinsic dustiness in seven different bulk soya consignments recently imported into the UK and Ireland and extended to include the two major soya respiratory allergens (HSP and STI) in the generated dust fractions during dustiness testing. However, care has to be taken about not over-interpreting such results as defining actual worker exposure [96] but rather an indication of the relevant propensity of different bulks to generate dust and allergen aerosols of certain defined sizes.

The seven bulks tests included two pelletised hull and five meal bulks. None of them showed any visual fungal contamination. The mean concentration of allergens and endotoxin for meal and hull samples is shown in **Table 4**. Whereas the amount of extractable low molecular weight HSP in hull is 23-fold that in meal, there is also on average 4-fold more of 20 kDa STI in the hull product than the meal. Very low levels of the *Aspergillus fumigatus* allergen were found in all of the bulks. As reported previously [26], higher endotoxin levels tend to be found in hull than meal samples and may represent a potential additional respiratory risk [28, 85, 97].

Of the seven bulk samples, one sample showed "high" dustiness (gravimetric results) in both the thoracic and respiratory fractions compared with the other samples. This may suggest that this particular material may be the sort of bulk that produces small, buoyant dust particles that could travel further with prevailing winds and penetrate deep into the lungs to cause symptoms of irritation. Interestingly allergen concentrations in the smaller particles of this specific material did not parallel the gravimetric results. The levels of allergen in the three fractions sizes largely depend upon both the dust levels in those fractions and the amount of allergen that was readily extractable from the bulk material. So the highest concentration of HSP in small respirable particles was one of the two hull samples but generally of lower "dustiness". What is clear is that all the small respirable fractions of the seven generated dusts contained measureable but highly variable levels of allergen, and in 5/7 samples the HSP content was significantly greater than STI, even in meal samples. These data are consistent with asthma epidemics where there was distance between the point source and causation of asthma, the putative allergens being Gly m1 and Gly m 2, measured by the HSP immunoassay.

The data from the drum dustiness testing are compatible with a health risk for lung irritancy or allergic responses depending on the nature of the specific bulk material. The pelletised hull material (both of the tested hull products, **Figure 1**) seems to be largely assumed by harbour managers and importers to be a "low dust" product. However, it does show some evidence of breakdown after transportation and unloading and has higher content of low molecular weight allergens and endotoxins.

The third HSE study was an occupational hygiene survey in a different dock unloading soya, but the focus on this study was investigating the levels of airborne endotoxin, as well as the allergens STI and HSP (paper in preparation). The levels of these analytes in the unloaded

Type	STI (μg g^{-1})	HSP (μg g^{-1})	Asp f 1 (μg g^{-1})	Endotoxin (EU g^{-1})	Moisture (%)
Meal	127	122	17 × 10^{-3}	12,922[a]	8.3
	(28–270)	(54–196)	(5–33) × 10^{-3}	(1309–51,455)	(5.2–13.9)
Hull	528	2862	13 × 10^{-3}	66,577	7.4
	(258–798)	(2824–2900)	(8–19) × 10^{-3}	(52,769–80,364)	(6.1–8.7)

[a]Mean inflated by one high bulk value.

Table 4. Mean (range) of allergens and endotoxin extractable from the bulks and their moisture content.

bulks (hull and meal) were also measured. During both days a meal bulk was unloaded, while on the second day, a pelletised hull product was also handled that included manual cleaning or "trimming" of one of the ship's hold.

The dockside operation was very similar to the previous UK study. Bulk cargo ships were emptied using a dockside crane into a hopper, which was then used to load trailers and transported to the storage facility via a weighbridge. Inside the storage warehouse, the soya was tipped from the trailers and formed into piles, using a pusher vehicle loaded into lorries as required for onward transportation. Some lorries were also directly loaded from the hopper. In emptying the ship's hold, an excavator and Bobcat shovel loader were lowered into the hold allowing the grab crane to access material efficiently. Final "trimming" of the ship's hold was done by workers manually scraping and shovelling from the hold's sides. Spillages on the dockside were cleaned up manually and by the use of a shovel loader. Respiratory protective equipment was available to all staff and invariably worn by those working in the hold or as the hatch man, but not necessarily at other times. The excavator, crane, tractors, loading shovel, and pusher were all fitted with cab filtration.

As previously reported, higher levels of endotoxin and HSP were found in the hull bulk compared to the meal. Airborne sampling results showed geometric means (ranges) of airborne levels of dust, endotoxin, STI, and HSP during unloading of 1.6 (<1–62) mg m^{-3}, 34 (5–2450) EU m^{-3}, 146 (1–122,462) ng m^{-3}, and 608 (2–243,654) ng m^{-3}, respectively. Expressed as 8-hour TWAs, 29% of all personal samples and 100% of those involved in cleaning within the ship's hold had endotoxin levels greater than 90 EU m^{-3}, the limit for endotoxin proposed by the Netherlands [17]. All workers involved in trimming activity within the hold as part of their working day (both manually trimming and operating the excavator and Bobcat) had estimated endotoxin 8-hour TWAs of endotoxin, between 175 and 888 EU m^{-3}. Personal samples and static samples within the two vehicles involved in trimming activities suggested hold atmospheric levels of endotoxin between 275 and 2450 EU m^{-3}. Two other workers' exposure to endotoxin, when expressed as 8-hour TWAs and unrelated to trimming, breached the Dutch endotoxin guidance value. This happened on the second day, when hull was being unloaded, and is related to moving the bulk material within the storage warehouse. On day 2 when hull was being handled, there was twice as much endotoxin associated with the airborne dust collected compared with day 1.

So soya hull has higher levels of endotoxin associated with it, and considerable levels of airborne endotoxin are produced when it is handled and moved. A review of dust and endotoxin exposure in livestock farming suggested full-shift, average levels of inhalable dust and endotoxin between 0.8–10.8 mg m^{-3} and 300–6600 EU m^{-3}, respectively [16], and a review of grain dust exposure in the UK reported a geometric mean exposure levels for endotoxin and dust of 1150 EU m^{-3} and 4.4 mg m^{-3}, respectively. Trimming activities in the ship's hold appear consistent with this level of endotoxin exposure. Swan [98], in sampling cereal grain dust exposure on the ships during unloading at two UK docks, measured endotoxin levels between 59 and 190,000 EU m^{-3} for personal samples and 74,000–7.7 x 10^6 EU m^{-3} for static samples. Swan's study found an even more highly significant association between inhalable dust and endotoxin levels than what we found. These data may suggest higher endotoxin exposure from handling cereal grain dust in these circumstances.

Table 5 compares the airborne levels of dust (gravimetric), STI and HSP, in the two occupational hygiene studies that HSE has undertaken. The static samples at the periphery of the dockside operation, and beyond, in the first study have been removed to allow better comparison. The obvious high values in the upper ranges of the second dock study likely reflect that monitoring of trimming in the hold was monitored, an obviously dusty activity.

2.5. Controlling exposure to soya dust

The evidence from the Barcelona clearly shows the value of controlling emissions of soya dust during bulk soya unloading. Such measures decreased both the measured levels of airborne soya substantially and finally eliminated outbreaks of asthma epidemic that has been serious enough to cause fatalities [63]. However, the initial implementation of control measures in 1987 still led to further outbreaks in 1994 and 1996, and in 1998 the storage silos were fitted with even greater particle retaining filters. The value of airborne monitoring of soya aeroallergens, which started in Barcelona in 1986, was also shown [56].

A number of international authoritative and regulatory bodies have recognised the health hazards from grain dust, and, while soya is not a cereal, some have explicitly encompassed soya within their definition of grain dust [99] or highlighted the similarities in the hazards (intrinsic and extrinsic) posed by dust from grain and soya [100]. A number of regulatory and authoritative bodies in the USA, Canada, and Europe have set occupational exposure

	Gravimetric dust (mg m^{-3})	STI (ng m^{-3})	HSP (ng m^{-3})
First UK dock study [19]			
Personal samples	2.0 (1.2–4.5)	130 (33–3071)	583 (170–12,629)
Workplace static samples	0.7 (0.1–5.2)	216 (11–845)	1970 (40–7438)
Second UK dock study			
Personal samples	1.8 (0.04–62.3)	178 (5–122,463)	763 (15–243,654)
Workplace static samples	1.1 (0.2–35.6)	85 (1–69,956)	318 (2–139,390)

Table 5. Geometric means (range) of airborne concentrations of gravimetric dust and soya allergens compared in the two UK dock studies. Samples are categorised as personal samples or static/background samples.

limits for grain dust ranging between 1.5 and 10 mg m^{-3} 8 hour. TWA. While Great Britain has an exposure limit for grain dust of 10 mg m^{-3} (gravimetric measurement), this is augmented by the need under the Control of Substances Hazardous to Health (COSHH) Regulations, given soya is a respiratory sensitiser, to undertake risk assessments, control soya exposures to as low as reasonable practicable, and implement appropriate health surveillance.

Monitoring by gravimetric dust is not necessarily a good surrogate of the extent of exposure to soya allergens or endotoxin. Our two dock studies [19] [paper in preparation] identified that gravimetric measurements only explained 50–70% of the variation in the airborne levels of the two allergens measured, even in a relatively constrained number of bulks. The lack of a good relationship found between gravimetric dust and allergen (HSP and STI) levels in the respiratory-sized fraction generated by drum dustiness tests confirms this [18]. HSE's second dock study suggested that gravimetric dust measurements only explained about 29–57% of the variation in endotoxin levels over the 2 days of sampling.

Issues about health risks in bakeries from exposure to cereal flour dust have been extensively investigated [6, 70, 81, 101–103], with many solutions identified for reducing bakers' exposure in terms of engineering control, local exhaust ventilation, work activity modifications, and training. Such measures to reduce airborne levels of flour should also reduce exposure to soya flour dust. Interestingly one suggestion for reducing the dustiness of certain flours, such as in improver mixes, has been the addition of soya oil [79].

3. Conclusions

The importance of soya in the global nutrition of animals and humans is well recognised. Largely cultivated in the USA or South American countries, it involves large-scale handling, processing, transportation, and finally use by a wide variety of end-users. Health problems from exposure to soya dust have been found in those directly occupationally exposed and those in the general population, indirectly exposed from occupational/agricultural activities. The major health problem seems to relate to type I, sIgE-mediated allergic reactions. There appears to be a genetic component to sensitisation; atopy status and exposure to soya dust are both significant risk factors, as well as smoking. Interestingly the reports of "flu-like" symptoms, similar to ODTS or HP, in two studies of soya processing suggest that an additional pathological mechanism can occur.

The UK largely imports soya meal and soya hull; some meal products also have an amount of hull deliberately added. The protein and allergen profiles of the two pure products are very different. Hull, which is used solely as an animal feedstuff, has a particular low molecular weight protein signature, including the two allergens Gly m 1 and Gly m 2 identified as causative in harbour city asthma epidemics. Generally, energetic handling of hull at ports can lead to high airborne concentrations of dust containing these allergens and that can travel distances up to at least 200 m in the direction of the prevailing wind. On a smaller scale, energetic handling of hull-based animal feed may produce considerable airborne levels of allergen and endotoxin. Limited data suggest that soya products can be very different in their propensity to be dusty, the particle sizes generated, and their allergen content.

Methods for monitoring airborne levels of relevant soya allergens are available and can be used to good effect in monitoring the efficacy of control measures.

Hull products appear to have a higher endotoxin load that can be become airborne: endotoxin posing its own respiratory risks. Also poor storage conditions can lead to significant growth of fungal contamination, some of the fungal species also being associated with respiratory ill health.

Those employed and living near large-scale operations of agriculture growing soya, storing, processing, and transportation may be exposed. Occupations such as stevedores, farmers, millers, bakers, and food processors may be exposed to soya dust. Bakers, which have had significant problems with occupational asthma and allergic symptoms from cereal flour, are likely to benefit from the measures enacted to control exposure to cereal flour dust, in reducing soya exposure to soya.

Regulatory regimes that tackle issues of respiratory problems from exposure to grain dust appear to either directly or indirectly encompass soya dust. Such measures may involve setting gravimetric workplace exposure limits, although the relationship between airborne dust levels and their allergen content is not necessarily simple. Great Britain has the further regulation of COSHH for asthmagens such as soya. This mandates employers to undertake risk assessments, keep exposure to as low as reasonable practicable, and utilise appropriate health surveillance. However, the efficacy of such a regulatory framework obviously depends on its implementation where soya is encountered.

Acknowledgements

This publication and some of the work it describes were funded by the Health and Safety Executive (HSE). Its contents, including any opinions and/or conclusion expressed, are those of the author alone and do not necessarily reflect HSE policy. My thanks to Andrew Simpson and Peter Baldwin for supplying photos of soya unloading in the UK.

Conflict of interest

The author declares no "conflict of interest".

Author details

Howard J. Mason

Address all correspondence to: howard.mason@hse.gov.uk

Health and Safety Executive, The Science and Research Centre, Buxton, UK

References

[1] Statistica Research Department. Soy Food Products Industry—Statistics & Facts. 2018. Available from: https://www.statista.com/topics/2218/soy-food-products-industry-statistics-and-facts/:Statistica

[2] Efeca. UK Roundtable on Sustainable Soya: Baseline Study 2018. Available from: https://www.efeca.com/the-uk-roundtable-on-sustainable-soya/2018

[3] Regulation (Eu) No 1169/2011 of the European Parliament and of the Council, 2011.

[4] Taylor SL, Nordlee JA, Sicherer SH, Sampson HA, Levy MB, Steinmand H, et al. Soybean oil is not allergenic to soybean-allergic individuals. Allergy and Clinical Immunology. 2004;**113**(Suppl 2):S99

[5] International Programme on Chemical Safety. Peanut Oils and Soya Bean Oils. Geneva: World Health Organisation; 2000

[6] van Tongeren M, Galea K, Ticker J, While D, Kromhout H, Cherrie J. Temporal trends of flour dust exposure in the United Kingdom, 1985-2003. Journal of Environmental Monitoring. 2009;**11**(8):1492-1497

[7] Baatjies R, Jeebhay M. Baker's allergy and asthma—a review of the literature. Current Allergy & Clinical Immunology. 2013;**26**(4):232-243

[8] Bush R, Schoeckenstein D, Meier-Davis S. Soybean flour asthma: detection of allergens by immunoblotting. The Journal of Allergy and Clinical Immunology. 1988;**82**:251-255

[9] Breiteneder H, Radauer C. A classification of plant food allergens. The Journal of Allergy and Clinical Immunology. 2004;**113**:821-830

[10] Hoffmann-Sommergruber R. Food allergen protein families and their structural characteristics: new data from Europrevail. Clinical and Translational Allergy. 2011;**1**(Suppl 1): S56

[11] Amnuaycheewa P. Gonzalez de Mejia E. Purification, characterisation and quantification of the soy allergen profilin (Gly m 3) in soy products. Food Chemistry. 2010;**119**:1671-1680

[12] L'Hocine L, Boye J. Allergenicity of soybean: New developments in identification of allergenic proteins, cross-reactivities and hypoallergenization technologies. Critical Reviews in Food Science and Nutrition. 2007;**47**:127-143

[13] Farokhi A, Heederik D, Smit L. Respiratory health effects of exposure to low levels of airborne endotoxin—a systematic review. Environmental Health. 2018;**17**:14

[14] Schwartz DA, Thorne PS, Yagla SJ, Burmeister LF, Olenchock SA, Watt JL, et al. The role of endotoxin in grain dust-induced lung disease. American Journal of Respiratory and Critical Care Medicine. 1995;**152**(2):603-608

[15] Smid T, Heederik D, Houba R, Quanjer PH. Dust- and Endotoxin-related Respiratory Effects in the Animal Feed Industry. The American Review of Respiratory Disease. 1992;**146**(6):1474-1479

[16] Basinas I, Sigsgaard T, Kromhout H, Heederik D, Wouters I, Schlunssen V. A comprehensive review of levels and determinants of personal exposure to dust and endotoxin in livestock farming. Journal of Exposure Science & Environmental Epidemiology. 2015;**25**:123-137

[17] Dutch Expert Committee on Occupational Standards. Endotoxins: Health Based Recommended Occupational Exposure Limit. The Netherlands: Health Council of The Netherlands; 2010

[18] Mason H, Gómez-Ollés S, Cruz M-J, Roberts P, Thorpe A, Evans G. Quantifying Dustiness, Specific Allergens and Endotoxin in Bulk Soya Imports. Environments. 2017;**4**(4):76

[19] Mason H, Gómez-Ollés S, Cruz M-J, Smith I, Evans G, Simpson A, et al. Levels of soya aeroallergens during dockside unloading as measured by personal and static sampling. Archives of Industrial Hygiene and Toxicology. 2015;**66**:23-29

[20] Heederik D, Doekes G, van Strien R, Brunekreef B. Daily changes of peak expiratory flow and respiratory symptom occurrence around a soy processing factory. Annals of Agricultural and Environmental Medicine. 2014;**21**(1):5-10

[21] Spies A, Rees D, Fourie A, Wilson K, Harris-Roberts J, Robinson E. Inhalable dust and protein exposure in soybean processing plants. International Journal of Occupational and Environmental Health. 2008;**14**(3):187-192

[22] Bohn J, BeMiller J. (1→3)-β-d-glucans as biological response modifiers. A review of structure-functional activity relationships. Carbohydrate Polymers. 1995;**28**:3-14

[23] Mousavi B, Hedayati M, Hedayati N, Ilkit M, Syedmousavi S. *Aspergillus* species in indoor environments and their possible occupational and public health hazards. Current Medical Mycology. 2016;**2**(1):36-42

[24] Codina R, Lockey R. Possible role of molds as secondary etiologic agents of the asthma epidemics in Barcelona, Spain. Journal of Allergy and Clinical Immunology. 1998;**102**(3):318-332

[25] Smith T, Lumley K, Hui E. Allergy to flour and fungal amylase in bakery workers. Occupational Medicine. 1997;**47**(1):21-24

[26] Harris-Roberts J, Robinson E, Fishwick D, Fourie A, Rees D, Spies A, et al. Sensitization and symptoms associated with soybean exposure in processing plants in South Africa. American Journal of Industrial Medicine. 2012;**55**:458-464

[27] Zuskin E, Kanceljak B, Schachter E, Witek T, Marom Z, Goswami S, et al. Immunological and respiratory changes in soy bean workers. International Archives of Occupational and Environmental Health. 1991;**63**(1):15-20

[28] Seifert S, Von Essen S, Jacobitz K, Crouch R, Lintner C. Organic dust toxic syndrome: a review. Journal of Toxicology. Clinical Toxicology. 2003;**41**:185-193

[29] Horner W, Helbling A, Salvaggio J, Lehrer S. Fungal allergens. Clinical Microbiology Reviews. 1995;**8**(2):161-179

[30] Selb R, Wal J, Moreno F, Lovik M, Mills C, Hoffmann-Sommergruber K, et al. Assessment of endogenous allergenicity of genetically modified plants exemplified by soybean – where do we stand? Food and Chemical Toxicology. 2017;**101**:139-148

[31] Geng T, Stojšin D, Liu K, Schaalje B, Postin C. J W, et al. Natural Variability of Allergen Levels in Conventional Soybeans: Assessing Variation across North and South America from Five Production Years. Journal of Agricultural and Food Chemistry. 2017;**18**(65):463-472

[32] González R, Zapatero L, Caravaca F, Carreira J. Identification of soybean proteins responsible for respiratory allergies. International Archives of Allergy and Applied Immunology. 1991;**95**(1):53-57

[33] González R, Polo F, Zapatero L, Caravaca F, Carreira J. Purification and characterization of major inhalant allergens from soybean hulls. Clinical and Experimental Allergy. 1992;**22**(8):748-755

[34] Codina R, Lockey R, Fernández-Caldas E, Rama R. Purification and characterization of a soybean hull allergen responsible for the Barcelona asthma outbreaks. II. Purification and sequencing of the Gly m 2 allergen. Clinical and Experimental Allergy. 1997;**27**(4):424-430

[35] Rihs H, Chen Z, Ruëff F, Petersen A, Rozynek P, Heimann H, et al. IgE binding of the recombinant allergen soybean profilin (rGly m 3) is mediated by conformational epitopes. The Journal of Allergy and Clinical Immunology. 1999;**104**(6):1293-1301

[36] Mittag D, Vieths S, Vogel L, Becker W, Rihs H, Helbling A, et al. Soybean allergy in patients allergic to birch pollen: clinical investigation and molecular characterization of allergens. The Journal of Allergy and Clinical Immunology. 2004;**113**(1):148-154

[37] Klemans R, Knol E, Michelsen-Huisman A, Pasmans S, de Kruijf-Broekman W, Bruijnzeel-Koomen C, et al. Components in soy allergy diagnostics: Gly m 2S albumin has the best diagnostic value in adults. Allergy. 2013;**68**(11):1396-1402

[38] Xiang P, Baird L, Jung R, Zeece M, Markwell J, Sarath G. P39, a Novel Soybean Protein Allergen, Belongs to a Plant-Specific Protein Family and Is Present in Protein Storage Vacuoles. Journal of Agricultural and Food Chemistry. 2008;**56**(6):2266-2272

[39] Xiang P, Haas E, Zeece M, Markwell J, Sarath G. C-Terminal 23 kDa polypeptide of soybean Gly m Bd 28 K is a potential allergen. Planta. 2004;**220**(1):56-63

[40] Ogawa T, Tsuji H, Bando N, Kitamura K, Zhu Y, Hirano H, et al. Identification of the soybean allergenic protein, Gly m Bd 30K, with the soybean seed 34-kDa oil-body-associated protein. Bioscience, Biotechnology, and Biochemistry. 1993;**57**(6):1030-1033

[41] Baur X, Pau M, Czuppon A, Fruhmann G. Characterisation of soybean allergens causing sensitisation of occupationally exposed bakers. Allergy. 1996;**51**:326-330

[42] Holzhauser T, Wackermann O, Ballmer-Weber B, Bindslev-Jensen C, Scibilia J, Perono-Garoffo L, et al. Soybean (Glycine max) allergy in Europe: Gly m 5 (b-conglycinin) and Gly m 6 (glycinin) are potential diagnostic markers for severe allergic reactions to soy. The Journal of Allergy and Clinical Immunology. 2009;**123**:452-458

[43] Riascos J, Weissinger S, Weissinger A, Kulis M, Burks A, Pons L. The Seed Biotinylated Protein of Soybean (Glycine max): A Boiling-Resistant New Allergen (Gly m 7) with the Capacity To Induce IgE-Mediated Allergic Responses. Journal of Agricultural and Food Chemistry. 2016;**64**(19):3890-3900

[44] Tarlo S, Lemiere C. Occupational Asthma. The New England Journal of Medicine. 2014;**370**(7):640-649

[45] Duke W. Soybean as a possible important source of allergy. Allergy. 1934;**5**:300-302

[46] Bush R, Cohen M. Immediate and late onset asthma from occupational exposure to soybean dust. Clinical and Experimental Allergy. 1977;**7**(4):369-373

[47] Pepys J. Occupational allergic lung diseases caused by organic agents. The Journal of Allergy and Clinical Immunology. 1986;**5**:1058-1062

[48] Cummings K, Gaughan D, Kullman G, Beezhold D, Green B, Blachere F, et al. Adverse respiratory outcomes associated with occupational exposures at a soy processing plant. The European Respiratory Journal. 2010;**36**:1007-1015

[49] White M, Etzel R, Olsen D, Goldstein I. Re-examination of epidemic asthma in New Orleans in relation of the presence of soy at the harbour. American Journal of Epidemiology. 1997;**145**:432-438

[50] Navarro C, Márquez M, Hernando L, Galvañ F, Zapatero L, Caravaca F. Epidemia asthma in Cartagena, Spain and its association with soybean sensitivity. Epidemiology and Community Health. 1993;**4**:76-79

[51] Hernando L, Navarro N, Marquez M, Zapatero L, Galvan F. Asthma epidemics and soybean in Cartagena (Spain). Lancet. 1989;**333**(8636):502

[52] Rovira E, Cuadras A, Gaig P, Gázquez V, Dalmau G, Gómez-Ollés S, et al. Soybean hull unloading in Tarragona (Spain) and asthma outbreak risk [Article in Spanish]. Gaceta Sanitaria. 2010;**24**(2):109-114

[53] Albert R. A propos de douze cas de sensibilisation au groupe des légumineuses comestibles [Twelve cases of sensitization to the group of edible leguminosa, in French]. Revue Française d'Allergologie. 1973;**13**:399-410

[54] Cocco G, Schiano M, Sacerdote G, Sagliocca L. Functional characteristics in soybean asthma. American Journal of Respiratory and Critical Care Medicine. 1995;**152**(Suppl):469

[55] Ballester F, Soriano J, Otero I, et al. Asthma visits to emergency rooms and soybean unloading in the harbours of Valencia and Coruna. American Journal of Epidemiology. 1999;**149**:315-322

[56] Aceves M, Grimalt J, Sunyer J, Anto J, Reed C. Identification of soybean dust as an epidemic asthma agent in urban areas by molecular marker and RAST analysis of aerosols. The Journal of Allergy and Clinical Immunology. 1991;**88**(1):124-134

[57] Carroll R. Epidemiology of New Orleans epidemic asthma. American Journal of Public Health. 1968;**58**:1677-1683

[58] Swanson M, Li J, Wentz-Murtha P, Trudeau W, Fernandez-Caldas E, Greife A, et al. Source of the aeroallergen of soybean dust: A low molecular mass glycopeptide from the soybean tela. The Journal of Allergy and Clinical Immunology. 1991;**87**:733-738

[59] Soriano J, Ercilla G, Sunyer J, Real F, Lázaro C, Rodrigo M, et al. HLA Class II Genes in Soybean Epidemic Asthma Patients. American Journal of Respiratory and Critical Care Medicine. 1997;**156**:1394-1398

[60] Anto J, Sunyer J. Epidemiologic studies of asthma epidemics in Barcelona. Chest. 1990;**96**(5):185S-190S

[61] Anto J, Sunyer J, Rodriguez-Roisin R, Suarez-Cervera M, Vasquez L. Community outbreaks of asthma associated with inhalation of soybean dust. The New England Journal of Medicine. 1989;**320**:1097-1102

[62] Antó J, Sunyer J, Newman TA. Comparison of soybean epidemic asthma and occupational asthma. Thorax. 1996;**51**(7):743-749

[63] Rodrigo M, Cruz M-J, Garcia M, Anto J, Genover T, Morell F. Epidemic asthma in Barcelona: An evaluation of new strategies for the control of soybean dust emission. International Archives of Allergy and Immunology. 2004;**134**:158-164

[64] Gonzalez R, Varela J, Carreira J, Polo F. Soybean hydrophobic protein and soybean hull allergy. Lancet. 1995;**346**:48-49

[65] Cruz M, Rodrigo M, Anto J, Morell F. An amplified ELISA inhibition method for the measurement of airborne soybean allergens. International Archives of Allergy and Immunology. 2000;**122**:42-48

[66] Gómez-Ollés S, Cruz M, Renström A, Doekes G, Morell F, Rodrigo M. An amplified sandwich EIA for the measurement of soy aeroallergens. Clinical and Experimental Allergy. 2006;**36**(9):1176-1183

[67] Gonzalez R, Duffort O, Calabozo B, Barber D, Carreira J, Polo F. Monoclonal antibody-based method to quantify Gly m 1. Its appliction to assess environmental exposure to soybean dust. Allergy. 2000;**55**:59-64

[68] Antonicelli L, Ruello M, Monsalve R, Gonzalez R, Fava G, Bonifazi F. Assessment of airborne soy-hull allergen (Gly m 1) in the Port of Ancona, Italy. European Annals of Allergy and Clinical Immunology. 2010;**42**(5):178-185

[69] Sunyer J, Anto J, Rodrigo M, Roca J, Morell F. Risk factors of soybean epidemic asthma : the role of smoking and atopy. The American Review of Respiratory Disease. 1992;**145**:1096-1102

[70] Harris-Roberts J, Robinson E, Waterhouse J, Billings C, Proctor A, Stocks-Greaves M, et al. Sensitization to wheat flour and enzymes and associated respiratory symptoms in British bakers. American Journal of Industrial Medicine. 2009;**52**(2):133-140

[71] Cullinan P, Lowson D, Nieuwenhuijsen M, Gordon S, Tee R, Venables K, et al. Work related symptoms, sensitisation, and estimated exposure in workers not previously exposed to laboratory rats. Occupational and Environmental Medicine. 1994;**51**(9):589-592

[72] Jeebhay M, Cartier A. Seafood workers and respiratory disease: an update. Current Opinion in Allergy and Clinical Immunology. 2010;**10**:104-113

[73] McSharry C, Anderson K, McKay I, Colloff M, Feyerabend C, Wilson R, et al. The IgE and IgG antibody responses to aerosols of Nephrops norvegicus (prawn) antigens: the association with clinical hypersensitivity and with cigarette smoking. Clinical and Experimental Allergy. 1994;**97**(3):499-504

[74] Alvarez M, Tabar A, Quirce S, Olaguíbel J, Lizaso M, Echechipía S, et al. Diversity of allergens causing occupational asthma among cereal workers as demonstrated by exposure procedures. Clinical and Experimental Allergy. 1996;**26**(2):147-153

[75] Patouchas D, Sampsonas F, Papantrinopoulou D, Tsoukalas P, Karkoulias K, Spiropoulos K. Determinants of specific sensitization in flour allergens in workers in bakeries with use of skin prick tests. European Review for Medical and Pharmacological Sciences. 2009;**13**:407-411

[76] Cannon J, Jones M, Welch J, Fitzgerald M, Szram J, Cullinan P. Prevalence of sensitisation to soya flour in the baking industry within the uk. Thorax. 2014;**69**(Suppl 2):A138

[77] Baur X, Sauer W, Weiss W. Baking Additives as New Allergens in Baker's Asthma. Respiration. 1988;**54**:70-72

[78] Lavaud F, Perdu D, Prévost A, Vallerand H, Cossart C, Passemard F. Baker's asthma related to soybean lecithin exposure. Allergy. 1994;**49**:159-162

[79] Mason H, Fraser S, Thorpe A, Roberts P, Evans G. Reducing dust and allergen exposure in bakeries. AIMS Allergy and Immunology. 2017;**1**:4194-4206

[80] Salcedo G, Quirce S, Diaz-Perales A. Wheat Allergens Associated With Baker's Asthma. Journal of Investigational Allergology & Clinical Immunology. 2011;**21**(2):81-92

[81] Baur X, Posch A. Characterized allergens causing bakers' asthma. Allergy. 1998;**53**(6): 562-566

[82] Quirce S, Polo F, Figueredo E, Gonzalez R, Sastre J, et al. Clinical and Experimental Allergy. 2000;**30**:839-846

[83] Zuskin E, Skuric Z, Kancelja B, Pokrajac D, Schachter N, Witek T. Respiratory Symptoms and Ventilatory Capacity in Soy Bean Workers. American Journal of Industrial Medicine. 1988;**14**(2):157-165

[84] Cummings K, Gaughan D, Green B. Flu like illness among workers at a soy processing plant. American Journal of Industrial Medicine. 2013;**56**:520-521

[85] Kirkhorn S, Garry V. Agricultural lung diseases. Environmental Health Perspectives. 2000;**108**(Suppl 4):705-712

[86] Zubeldia J, Gil P, Miralles P, de Barrio M, Aranzábal A, Herrero T, et al. Hypersensitivity pneumonitis caused by soybean antigens. The Journal of Allergy and Clinical Immunology. 1995;**95**(2):622-626

[87] Green B, Cummings K, Rittenour W, Hettick J, Bledsoe T, Blachere F, et al. Occupational sensitisation to soy allergens in workers at a processing facility. Clinical and Experimental Allergy. 2011;**41**(7):1022-1030

[88] Maggio P, Monsó E, Baltasar M, Morera J. Occupational asthma caused by soybean hull: a workplace equivalent to epidemic asthma. Allergy. 2003;**58**(4):350-351

[89] Heederik D, Doekes G, van Strien R, Brunekreef B. Daily changes of peak expiratory flow and respiratory symptom occurence around a soy processing factory. Annals of Agricultural and Environmental Medicine. 2014;**21**(1):5-10

[90] Codina R, Ardusso L, Lockey R, Crisci C, Bertoya N. Sensitisation to soybean hull allergens in sibjects exposed to different levels of soybean dust inhalation in Argentina. The Journal of Allergy and Clinical Immunology. 2000;**105**:570-576

[91] Odani S, Koide T, Ono T, Seto Y, Tanaka T. Soybean hydrophobic protein: isolation, partial characterisation and the complete primary structure. European Journal of Biochemistry. 1987;**162**:485-491

[92] Simpson A, Mason H. Inhalable Soya Dust Exposure at xxxxx Dock. Contract No.: OH/LET/2011/FSA/68. Buxton: HSL; 2011

[93] Health and Safety Executive. Dustiness of Powders and Materials. Sudbury, UK: HSE Books; 1996

[94] European Committee for Standardization. Workplace Atmospheres—Measurement of the Dustiness of Bulk Materials—Requirements and Test Methods. EN 15051. Brussels, Belgium: CEN; 2006

[95] Brouwer D, Links I, De Vreede S, Christopher Y. Size selective dustiness and exposure: simulated workplace comparisons. Annals of Occupational Hygiene. 2006;**50**(5):445-452

[96] Heitbrink W, Todd W, Cooper T, O'Brien D. The application of dustiness tests to the predictio of worker dust exposure. American Journal of Industrial Medicine. 1990;**51**:217-223

[97] Von Essen S, Fryzek J, Nowakowski B, Wampler M. Respiratory symptoms and farming practices in farmers associated with an acute febrile illness after organic dust exposure. Chest. 1999;**116**:1452-1458

[98] Swan J, Blainey D, Crook B. The HSE Grain Study-Workers' Exposure to Grain Dust Contaminants, Immunological and Clinical Response. RR 540. Health and Safety Executive, 2007

[99] Dutch Expert Committee on Occupational Standards. Grain Dust: Health-based recommended occupational exposure limit. The Hague: A Committee of the Health Council or the Netherlands; 2011

[100] Health and Safety Executive. Grain Dust: Guidance Note (Third edition) EH66. HSE; 2013. Available from: www.hse.gov.uk/pubns/eh66.htm

[101] Stobnicka A, Gorny R. Exposure to flour dust in the occupational environment. International Journal of Occupational Safety and Ergonomics. 2015;**21**(3):241-249

[102] Wiley K, Smith M, Allan L, Griffin P. Measurement of airborne flour exposure with a monoclonal antibody-based immunoassay. International Archives of Allergy and Immunology. 1997;**114**:278-284

[103] Griffin P, Fishwick D, Elms J, Curran A. Respiratory symptoms and wheat flour exposure: a study of flour millers. Occupational Medicine. 2001;**51**:141-143

Chapter 5

The Potential Impacts of Soy Protein on Fish Gut Health

Vikas Kumar, Md. Sakhawat Hossain,
Janice A. Ragaza and Marina Rubio Benito

Additional information is available at the end of the chapter

http://dx.doi.org/10.5772/intechopen.92695

Abstract

Soy protein is the major source of protein as fishmeal replacement in fish feed because of its worldwide availability and low price. However, the presence of high carbohydrate content along with saponins, lectins, and phytates can have a negative impact on fish gut health. Based on the literature and our lab studies, dietary soybean meal can cause a dose-dependent type of distal intestine inflammation called enteritis in commercial fish species including salmonids. This leads to reduced absorptive capacity, increased mucus secretion, hyperpermeability, and leucocyte infiltration in the lamina propria and submucosa, also inducing the pro-inflammatory cytokine genes expression, including Il-1β, Il-8, and Tnf-α. In addition, dietary soy may alter the composition and population of the gut microbiota via providing nutrients and energy that preferentially support the growth of some gut bacteria. This chapter summarizes the current knowledge of the effects of soy protein on the enteritis and gut microbiota.

Keywords: aquaculture, fish feed, soy protein, growth performance, enteritis, microbiome

1. Introduction

Soybean meal (SBM) is one of the most commonly used alternative plant-based ingredients to replace marine derived fishmeal (FM) in aquafeed. Relatively, high protein content and favorable amino acid profile of SBM approaches the nutritional requirement of many cultured species [1–4]. In carnivorous fish species almost 20–40% fishmeal protein can be replaced by SBM protein without compromising growth, feed utilization performances, and gut health [5]. It is also well documented that high dietary soy protein inclusion resulted in lower feed intake, reduced weight gain, morphological changes of distal intestinal epithelium, and abnormal health condition of fish [6–12]. The challenges behind the high inclusion of soy protein in aquafeed includes the limiting amino acids methionine, presence of high carbohydrate level which negatively influences mainly carnivorous species, and the presence of different antinutritional factors (ANFs) [1–3]. To overcome the challenges, several techniques have been attempted viz. using different processing techniques (heat and enzymatic treatment, bioprocessing, fermentation, etc.) to improve the soy ingredient profiles like increased protein levels, decreased levels of ANFs, and enhanced digestibility [13–15]. For balancing amino acids profile, a balance mixture of soy protein with other plant ingredients protein and crystalline amino acids supplementation were also practiced. However, SBM of standard quality is used in carnivorous fish diets only at relatively low levels due to its negative effects on gut health in several fish species [16]. Different soy protein sources have been found to modulate many aspects of gastrointestinal tract (GIT) health within fish species, including the histological composition, immune status, and the overall intestinal microbiota [17–23]. Intestinal morphology, gut-associated immunity, and microbial community are closely interacting with each other. The present chapter addresses the potential impacts of different soy protein inclusion in aquafeed on gut health condition of fish with special emphasis on gut morphology, soybean meal-induced enteritis (SBMIE), gut-associated immunity, and gut microbiota.

2. Effect of soybean meal inclusion in aquafeed to induced enteritis (SBMIE) in fish

When studying enteritis in fish, it is important to consider all cell types involved in a correct function of the gastrointestinal tract (GIT). Like mammals, fish gut is critical for nutrient digestion and absorption, immunity, and interaction with the environment [24–26]. In fish, a simpler division of the GIT is described as compared to mammals: two different segments of the gut are distinguished: proximal or anterior intestine and distal or posterior intestine (**Figure 1a,b**). Of these, distal intestine is where most of the nutrient absorption occurs and is the object of study of SBMIE in fish [28–33].

The word enteritis refers to an inflammatory process happening in the gut, which can be caused by a diverse range of factors. The symptoms that define the condition are a shortening of the mucosal folds, a loss of the normal supranuclear vacuolization of the absorptive cells in the intestinal epithelium, a widening of the central stroma within the mucosal folding, with increased amounts of connective tissue, a profound infiltration of inflammatory cells in the lamina propria [4, 6, 19, 34–37], an increased presence of IgM [38], an increased amount of goblet cells in the epithelium, as well as a decreased height of the microvilli together with increased microvillar vesicle formation [36]. Baeverfjord and Krogdahl [6] also described this condition as "a non-infectious sub-acute inflammation of the distal intestine." Typically, FM replacement by SBM in fish diets is between 20 and 40% at which signs of enteritis are detected in a large variety of both marine and freshwater species including omnivores and

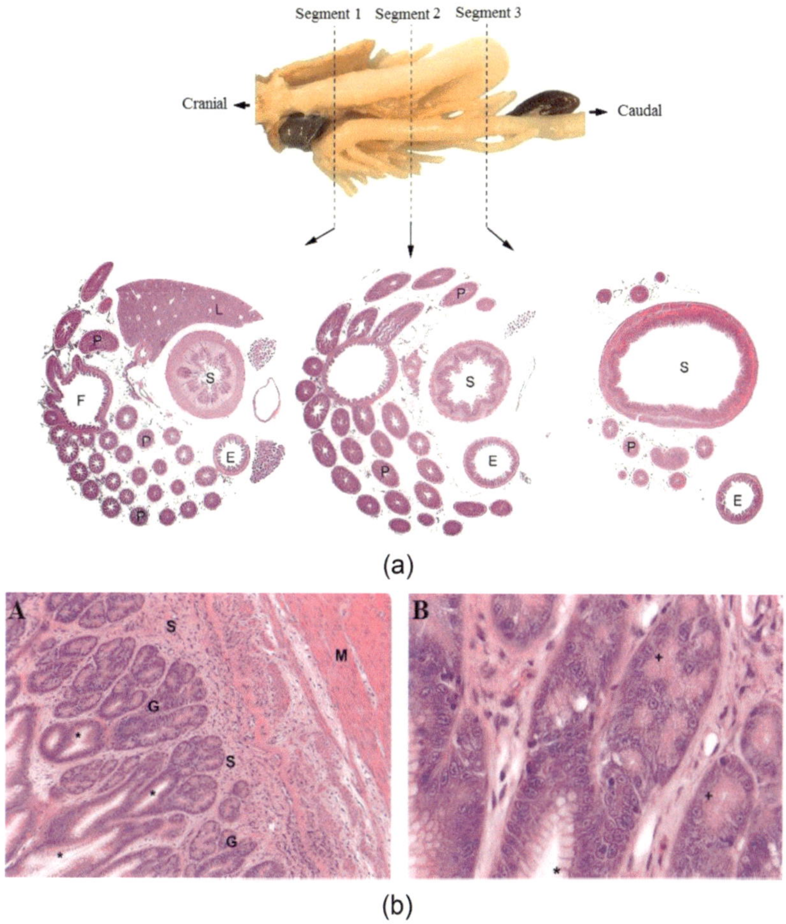

Figure 1. (a) On top: Different section-planes through the digestive tract of the rainbow trout. Below: Sections through segment 1, 2, and 3 consisting of Liver (L), stomach (S), foregut (F) with pyloric appendices (P), hind gut/rectum (E) (Source: [27]). (b) (A) rainbow trout stomach showing well-developed gastric glands (G), submucosa (S), and the tunica muscularis (M). (B) Higher magnification of branched tubular gastric glands. Asterisk (*) represents Shape and cellular morphology of mucous and (+) indicates pepsin and hydrochloric acid producing cell-type (oxyntopeptidic cells) (Source: [27]).

carnivores [39]. A reduction on feed intake and consequent decrease in weight gain is the first indicator that a given diet is exerting a negative effect.

There are several factors affecting the occurrence of SBMIE such as the inclusion levels, varieties, origins, and processing techniques of the different soybean products along with species variation and husbandry conditions (temperature, salinity, etc.). Nordrum et al. [40] who investigated the effect of salinity on the development of enteritis in salmonids. Urán [41] reported that with the increasing water temperature, the metabolic rate of Atlantic salmon increased which help to increase the severity of enteritis. Regarding the species variation effects on SBMIE, Nordrum et al. [40] also found that the effects of SBM on the intestinal morphology of rainbow trout were of less magnitude than for salmon. Similarly, Booman et al. [42] reported that soybean meal induced enteritis in Atlantic salmon (*Salmo salar*) and Chinook salmon (*Oncorhynchus tshawytscha*) but not in pink salmon (*O. gorbuscha*).

To understand the mechanism of intestinal inflammation or enteritis, it is important to also understand the elements involved at the cellular level. Enterocytes are cuboidal shaped epithelial cells that are distinct on their apical surface than on their basal surface. The apical surface faces the intestinal lumen and shows the characteristic folding of intestinal cells, called microvilli. The basal side is connected to vasculature where absorbed nutrients are released. Nutrients can undergo transcellular transport on the apical and basolateral membranes of the cell; this can happen by diffusion or by active transport through transmembrane transporters like the glucose/Na+ cotransport system, or other amino acid transporters like glutamine transporter, and also through pumps and channels. Paracellular transport occurs in between epithelial cells, and only small molecules and ions, solutes, and fluids can reach the blood this way. Only small nutrients can diffuse this way, as, in healthy conditions, enterocytes are held together through important tight junctions that keep the intestinal integrity [31, 32]. Intercellular exchange between enterocytes is also possible though gap junctions and desmosomes. **Figure 2** showing the healthy and abnormal enterocyte condition in fish.

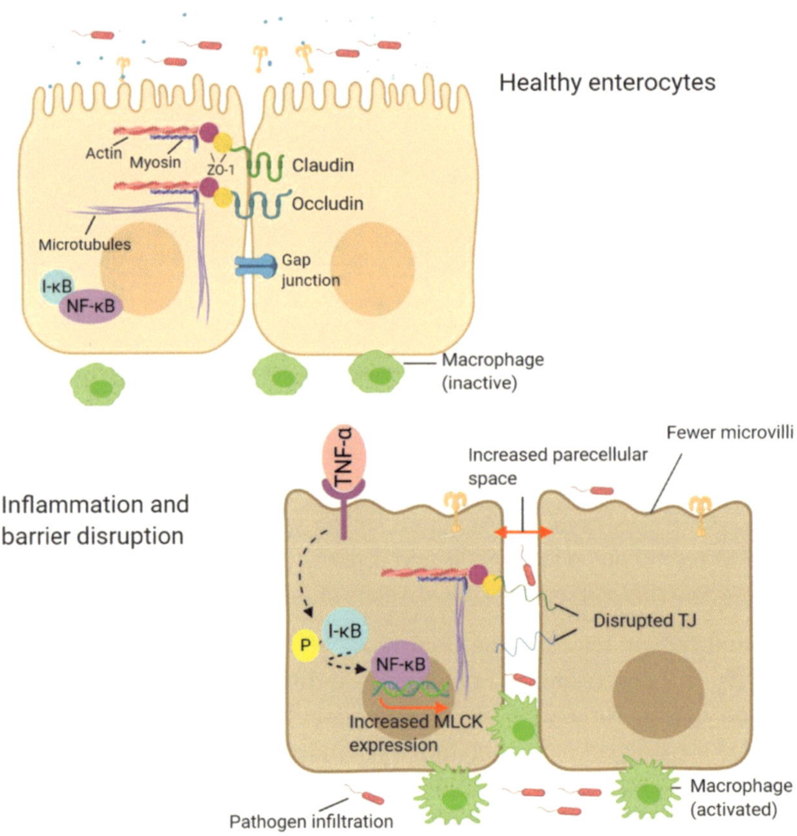

Figure 2. Figure showing the healthy and abnormal enterocyte condition in fish.

Some tight junction proteins that have been studied in fish with regards to SBMIE include transmembrane proteins like occludins and claudins and intracellular components like zonula occludens-1 (ZO-1) [26, 28, 31, 32]. Another intracellular component that interacts with the tight junction complex is myosin light chain kinase (MLCK), involved in cytoskeletal contraction, smooth muscle contraction, and, therefore, tight junction regulation and paracellular permeability [26, 43, 44].

3. Morphophysiological effects at fish gut of soy protein inclusion in aquafeed

Although soy protein has widely been used in aquafeed as a cheap alternative protein source for FM; however, the presence of some ANFs in SBM restricts its level of inclusion in aquafeed. High inclusion of soybean ingredients causes several negative effects on palatability and intestinal

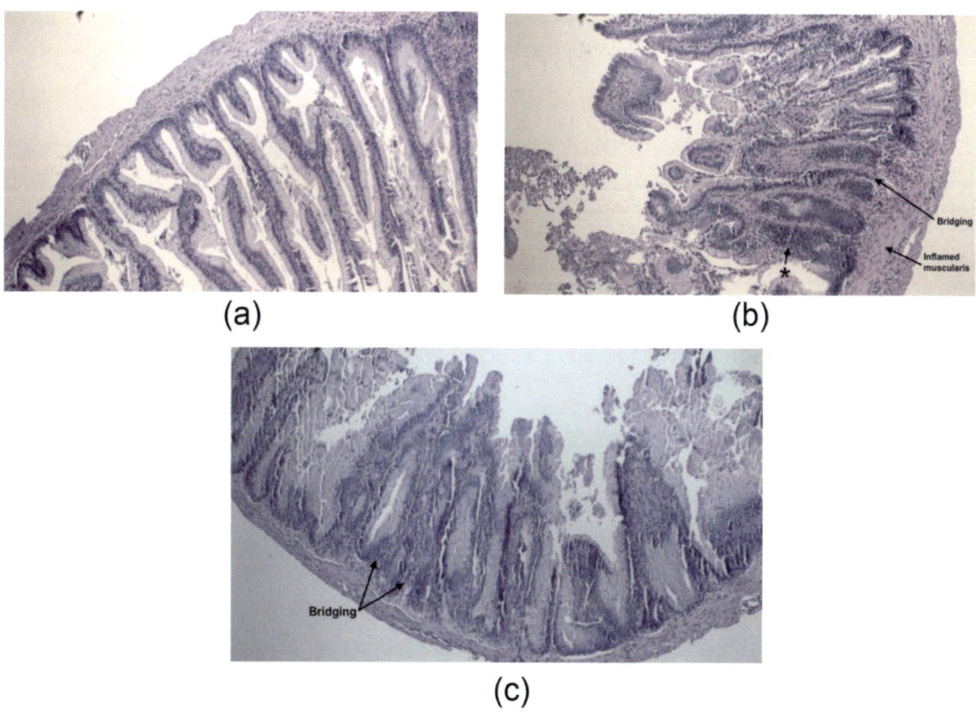

Figure 3. (a) Intestinal histology (posterior, intestine, 20 X) of rainbow trout (*O. mykiss*) fed FM-based control diet showing normal condition of intestine (Kumar et al., unpublished data). (b) Intestinal histology (posterior, intestine, 20 X) of rainbow trout (*O. mykiss*) fed low SBM (10.3%)-based diet showing inflamed muscularis, leukocyte infiltration of the lamina propria leading to swelling and mucosal fold fusion (bridging). Increased prevalence of globlet cells possibly to secrete more mucous to protect the epithelium. Asterisk denotes inflammation (Kumar et al., unpublished data). (c) Intestinal histology (posterior, intestine, 20 X) of rainbow trout (*O. mykiss*) fed high SBM (20.7%) based diet showing villi and lamina propria highly inflamed (leading to much wider mucosal folds), muscularis inflamed, villi shortened, disorganization of epithelium, reduction in supranuclear absorptive vacuoles, mucosal fold fusion (bridging) and some structural disintegration. By far, this treatment led to the most changes (Kumar et al., unpublished data).

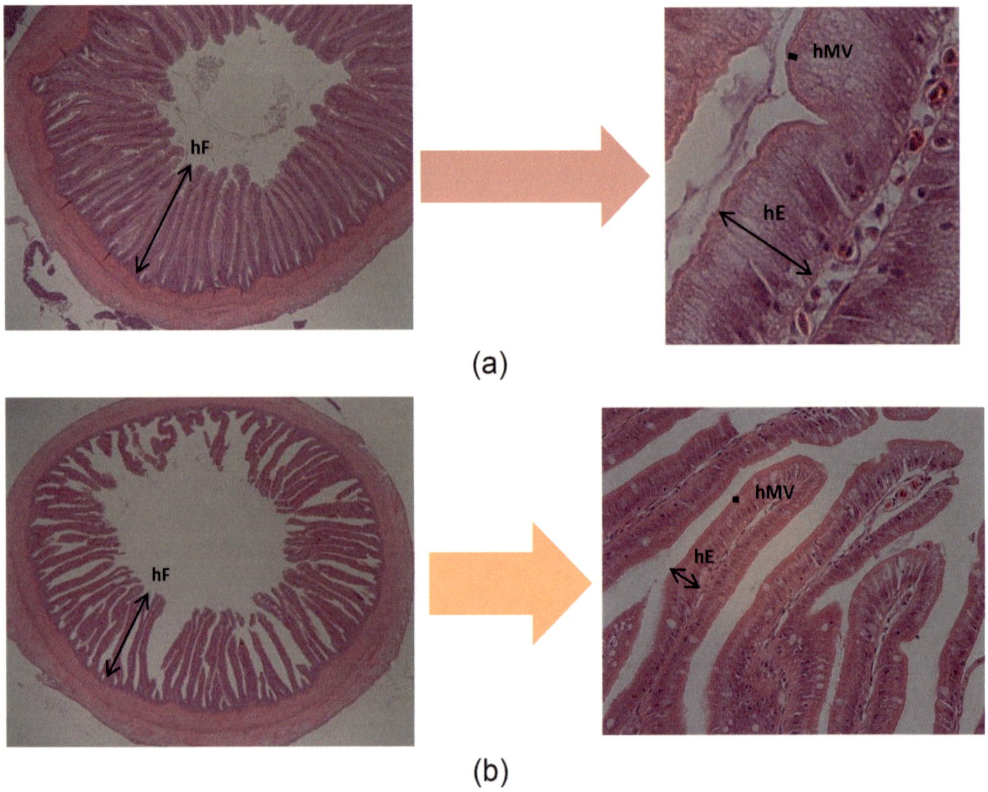

Figure 4. (a-1) Cross-section of anterior intestine (X 40, HE) of Amberjack (*Seriola dumerili*) fed FM-based control diet showing increased fold height (hF). (a-2) Cross-section of anterior intestine (X 400, HE) of Amberjack (*Seriola dumerili*) fed FM-based control diet showing increased enterocyte height (hE) and microvillus height (hMV). (b-1) Cross-section of anterior intestine (X 40, HE) of Amberjack (*Seriola dumerili*) fed 50% FM replaced with SBM-based diet showing reduced fold height (hF). (b-2) Cross-section of anterior intestine (X 400, HE) of Amberjack (*Seriola dumerili*) fed 50% FM replaced with SBM-based diet. Showing reduced enterocyte height (hE) and microvillus height (hMV).

morphology of fish. From previous researches, it is well documented that high inclusion of soybean meal (>40%) causes several intestinal morphological changes such as the reduction in mucosal folding, reduced fold height, enterocyte height, microvillus height, loss of mucosal integrity, abnormal vacuolization, and inflammatory cell infiltration (**Figures 3a–c** and **4a–1, a–2, b–1, b–2**) in aquatic animals [5–8, 36, 45–49]. The degree of morphological changes in the intestine depends on the inclusion level of SBM which is also correlated with the cultured fish species. Reduced fold height, enterocyte height, and microvillus height reduced the area of nutrient absorption in the intestine which finally affects the fish performances.

Feed nutrients must be digested for their utilization, and pancreatic digestive enzymes have essential roles for the digestion; trypsin and chymotrypsin are the main pancreatic proteases, lipase is the major pancreatic lipolytic enzyme, and amylase is known as the major pancreatic digestive enzyme for carbohydrates (Murashita et al. [50]). Inclusion of SBM that also affects the digestive enzyme secretion of different fish species is well documented. Murashita et al. [50] reported that red sea bream fed SBM showed lower content and activity of four pancreatic digestive enzymes compared to fish fed FM. Also, lower gene expression levels of the digestive enzymes in the hepatopancreas were observed in the SBM fed red sea bream compared with the FM fed fish, which is in line with the report in yellowtail; orally administrated FM increased the trypsin and lipase gene expressions in the pyloric caeca, but not in fish administrated SBM [51]. Perera and Yúfera [52] reported that early SBM feeding of *S. aurata* larvae significantly affects the activity of most pancreatic enzymes in a time-of-exposure dependent form. More than 10 days of SBM feeding (i.e., beyond 14 dph) delayed the normal development of *S. aurata* larvae digestive capacities as the activities of all trypsin, chymotrypsin, and amylase were significantly reduced. This is opposed to the typical response of juvenile and adult fish to SBM. Protease inhibitors present in SBM can partially abrogate the activity of trypsin and chymotrypsin in the proximal intestine [53], and juveniles of *S. aurata* [54] and other fish such as Atlantic salmon [55, 56] exhibit a rapid compensatory increase in activities of these enzymes. However, SBM-induced increase in trypsin activity in juvenile fish is more marked in the distal intestine and has been attributed to a reduced ability to reabsorb the pancreatic enzymes [11] and the upregulation of trypsin-like activity by immune cells [56]. Therefore, other plausible explanation for our observations is that these intestinal processes are not fully functional in early larvae. Conversely, lipase activity was relatively insensitive to SBM in *S. aurata* larvae, as reported before in postsmolt Atlantic salmon [55]. The observed decrease in pancreatic proteases may be responsible for less lipase inactivation in the digestive tract explaining the stable lipase activity.

4. Soy protein inclusion impacts on gut-associated immunity

The gastrointestinal tract carries many functions in teleost; among them, defense is possibly one of the most important functions. Gut acts as a physical barrier to pathogen entry that also contains a gut-associated lymphoid tissue (GALT). Teleost gut-associated lymphoid tissue (GALT) consists of leucocyte populations located both intraepithelially and in the lamina propria with no structural organization. The gut microbes play a critical role in the development and maturation of GALT, which in turn mediate a variety of host immune functions [49]. Recent work on the structure of GALT and other intestinal cell populations, the absorption of macromolecules from the intestinal lumen, and the production of specific mucosal antibodies strongly suggests, however, that the gut of these lower vertebrates is immunocompetent. Fish intestine, especially the posterior segment, is immunologically active and armored with various immune cell types, including B cells, macrophages, granulocytes, and T cells [57, 58]. Studies on the gut-associated immunity are important for the aquaculture industry for several reasons. First, the gut is one of the main portals of entry of pathogens. Second, farmed fish are generally fed commercial pellets, which give farmers the ability to manipulate fish health by incorporating drugs, vaccines, and different feed ingredients or additives into the feed. Third, the gut immune system of teleost's allows microbial colonization by symbionts, and this microbial community can be regarded as a mechanism to modulate fish pathogens [59]. Many studies attempt to reveal the effects of inclusion of plant origin ingredients and different feed additives on gut-associated immunity in fish. In this current chapter, an attempt has made to discuss the effects of soy protein ingredients inclusion in aquafeed on gut-associated immunity of various fish species.

Research conducted on the modulation of gut immune response due to the inclusion of soy products in aquafeed predominantly concentrate on innate immune parameters. Several molecules involved in innate immunity are found in the intestine of different fish species, such as lysozymes in Asian sea bass (*Lates calcarifer*) [60] and Atlantic salmon (*Salmo salar* L.) [61]; complement components in grass carp (*Ctenopharyngodon idella*) [62], rainbow trout (*Oncorhynchus mykiss*) [63, 64], and Asian sea bass [65]; cytokines in Atlantic cod (*Gadus morhua*) [66] and rainbow trout [67]; lectins in several species (reviewed in [68]); or antimicrobial peptides (AMPs) in rainbow trout [69] and grouper (*Epinephelus coioides*) [70].

The immune status of a fish's intestinal mucosa is closely associated with inflammation, which is mediated by cytokines. Cytokines, such as IL-1, IL-10, and IL-16 have a fundamental role in the regulation of inflammatory responses in fish throughout the infection process [31, 32]. Many studies have shown that IL-1 and IL-16 are increased in inflammatory bowel disease and there was a positive association between disease activity [71, 72]. While the deprivation of IL-10 evokes the development of inflammatory bowel disease, the decrease of IL-10 can aggravates local inflammation [73, 74]. Wang et al. [75] reported the increased expression levels of IL-1 and IL-16 mRNA with the SBM level in the diet, whereas the IL-10 mRNA expression level decreased with the SBM level in the diet of orange-spotted grouper (*Epinephelus coioides*). At the same time, infiltrate leucocytes were observed in the intestinal epithelium in grouper fed diets contained SBM. Furthermore, the degree of intestine inflammation was positively correlated with IL-1 and IL-16 mRNA expression levels but negatively correlated with expression of IL-10 mRNA. Their results suggest that SBM can cause intestinal inflammation by increasing pro-inflammatory cytokine levels and decreasing anti-inflammatory cytokine levels.

Krogdahl et al. [35] examined the effect of solvent and alcohol extracted SBM in the diets of Atlantic salmon, and their results indicated that fish fed solvent-extracted SBM showed higher mortality rate when challenged by *A. salmonicida*. In addition, fish fed alcohol-extracted SBM revealed increased levels of both lysozyme and IgM in the mid and distal intestinal mucosa. In Atlantic salmon, Lilleeng et al. [76] showed significantly downregulated TGF-β gene expression after on day feeding of extracted SBM (460 g kg^{-1}), whereas reduced expression of interferon-inducible lysosomal thiol reductase (GILT) was observed followed by 3 days feeding. The authors assumed that the downregulation of TGF-β and GILT might be due to the failure to maintain mucosal integrity in the distal intestine. Sahlmann et al. [77] investigated transcriptomic profiling in Atlantic salmon feeding of SBM (200 g kg^{-1}) containing diet for 1 week. On days 3 and 5, a prominent change in gene expression patterns was observed. Immune-related genes were upregulated during the first 5 days: GTPase IMAP family members; NF-kB-related genes; and regulators of T-cell and B-cell function. These immune genes expression profiles suggest that intestinal inflammation is induced within a week upon administration of an SBM-containing diet, which may in turn negatively influence the growth performance of salmonids.

Bruce et al. [29] evaluated processed soybean meal ingredients (defatted soybean meal, bio-processed soybean meal [BSBM], and commercial soy protein concentrate [CSPC]) inclusion in the diets of rainbow trout on intestinal immunity. They reported no significant differences in intestinal immunoglobulin concentrations ($p = 0.41$) or gut leukocyte phagocytosis at day 15 samplings ($p = 0.41$). Intestinal lysozyme activity showed some modulation throughout the feeding trial period, with the BSBM diet producing higher levels in the long-term sample (60 day). A previous study on Atlantic salmon (*Salmo salar*) showed increased lysozyme activity in the intestinal mucosa due to the dietary inclusion of soybean molasses, indicating a potential inflammatory response and was potential activation of leukocytes [35]. Kim and Austin [78] also found high lysozyme activity in rainbow trout intestinal mucus samples after the administration of probiotics compounds which may be closely related to bioprocessed plant-based ingredients. Therefore, increased lysozyme levels may also be indicative of intestinal innate immunity and gut health enhancements.

The mucosal immune system in fish includes certain immunocompetent cells and factors in the intestinal mucous membrane. Of these factors, the interleukins (ILs), interferon regulatory factors (IRFs), and tumor necrosis factors (TNFs) are the main immune-relevant factors linked to inflammation in the distal intestine in fish [79]. Recently Miao et al. [80] reported the substitution effects of dietary SBM on the mucosal immune system in northern snakehead through measuring the gene expression of certain inflammatory cytokines (IL-1β, IL-8, IL-10, and IL-17F) in the distal intestine. After 63-day feeding, trial results indicated that dietary soybean meal affected the gene expression of certain factors. The up-regulated relative expression of IL-1β in the fish fed diet group containing 75% defatted fishmeal replacement with SBM was consistent with the observations in Atlantic salmon [77, 81]. However, the level of IL-1β observed in the same diet group was only 1.6-fold higher than that in FM-based control diet, while that observed in Atlantic salmon was 20-fold higher [79]. The effect of dietary soybean meal on the expression of IL-1β reflects the fish species and stages due to the different tolerance capability for soybean meal [79].

5. Effects of soy protein inclusion on gut microbiota

Healthy gut microbiota is essential to promote host health and well-being. Before the 1970s, there were some controversies regarding the existence and role of an indigenous microbiota in fish. However, it is now well established that fish and other aquatic animals have a microbiota in the GI tract (for review, see; [21, 23, 82–92]). The intestinal microbiota of fish, as is the case of mammals, is classified as autochthonous (indigenous) or allochthonous bacteria [90, 93]. The autochthonous bacteria are those able to colonize the host's gut epithelial surface or are associated with the microvilli, while the allochthonous bacteria are incidental visitors in the GI tract and are expelled after some time without colonizing [90, 93]. Several factors affect the gut microbiota in fish including host factors, environmental factors, microbial factors, etc. However, until recently, among different influencing factors affecting the fish microbiota, water and diet (environmental factors) have been studied extensively [49]. In this section, we address the effect of dietary soybean products on intestinal bacterial community of finfish and crustaceans (**Table 1**).

5.1. In salmonids

Research conducted until recently on SBM inclusion effects on gut microbiota of fish indicated that SBM modulated the intestinal microbiota toward developing an undesirable microbial community that can induce mucosal inflammation [110, 111]. Heikkinen et al. [94] reported that rainbow trout fish fed FM- and SBM-based diets for 4 weeks showed decreased number

Species/initial weight	Soy protein type and feeding duration	Effects on gut microbiota	References
Rainbow trout			
21.1 ± 1.4 g	SBM (450 g/kg) for 8 weeks	↓ culturable bacteria, *Lactobacillus* spp., *Sphingomonas* spp. ↑ *Bacillus* spp., *Chryseomonas* spp.	Heikkinen et al. [94]
~40 g	SBM (450 g/kg) for 16 weeks	→ total culturable aerobic levels, *Micrococcus* spp. ↓ *Aeromonas* spp., *Vibrio* spp. ↑ *Actinomycetales*, *Psychrobacter* spp., *Saccharomyces* spp.	Merrifield et al. [95]
1.56 ± 0.9 kg	SBM (300 g/kg) for 8 weeks	↑ no. of clones identified as *Carnobacterium maltaromaticum* ↓ no. of different sequences in library	Mansfield et al. [96]
~510 g	SBM (300 g/kg) for 8 weeks	↑ *Firmicutes*: *Proteobacteria* ratio DGGE analysis revealed low similarity indices between SBM- fed fish and the control	Desai et al. [97]
17.21 ± 0.51 g	defatted soybean meal (SBM), bioprocessed soybean meal (BSBM) and commercial soy protein concentrate (CSPC) replaced approximately 73% menhaden fishmeal and fed for 60 days	The incorporation of processed soy-based proteins alters the microbial community composition within the distal intestine. Species diversity based on abundance and evenness were lowest in the SBM group, and were significantly less than the BSBM-L ($p = 0.003$) and CSPC ($p = 0.003$) treatments	[29]
Atlantic salmon			
172 g	SBM (250 g/kg) for 3 weeks	↑ autochthonous bacteria in MI and DI, allochthonous in DI → no. of genera and strains ↑ *Brevibacterium, Enterococcus,* yeast ↓ *Marinilactobiacillus psychrotolerans*, *C. maltaromaticum*	Bakke-McKellep et al. [98]
242 ± 8 g	SBM (436 g/kg) for 4 weeks	→ viable counts ↓ *Carnobacterium* spp., *Bacillus* spp.	Ringø et al. [99]
1204 ± 34 g	SPC (50 g/kg)	↑ bacterial diversity, *Escherichia coli,* a Pseudomonadales	Green et al. [100]
144.5 ± 2.3 g	SBM (378 g/kg) for 35 days	→ total and viable counts ↑ *Aeromonas* Via, *Sporosarcina equimarina*	Navarrete et al. [101]
305 ± 69 g	SPC (200 g/kg) for 12 weeks	→ total autochthonous bacteria (proximal Aintestine; PI), total allochthonous bacteria, allochthonous community composition and total autochthonous bacteria (DI) ↑ autochthonous *Enterobacteriaceae,* Bacilli-like, *Lactobacillaceae, Streptococcaceae* in PI ↓ autochthonous *Vibrionaceae* in PI ↑ autochthonous Bacilli- like, *Streptococcaceae* in DI	Hartviksen et al. [102]
~133 g	SBM (200 g/kg) for 80 days	↓ the diversity indices in DI, *Weissella confusa* in the DI, proportion of Photobacterium in MI ↑ relative abundance of *Firmicutes* compared with the FM group, abundance of *Lactococcus lactis* subsp. Lactis in MI → Photobacterium in DI	Reveco et al. [103]

Species/initial weight	Soy protein type and feeding duration	Effects on gut microbiota	References
	SBM (246 g/kg) for 84 days	↑ allochthonous bacterial level in FG, HG → allochthonous bacterial level in HC ↓ autochthonous bacterial level in FG, HG and HC	Refstie et al. [47]
	SBM (246 g/kg) for 84 days	Modulated gut microbiota. ↑ *Chryseobacterium* and *Psychrobacter*	Ringø et al., [22]
~534 g	BPSBM (214 g/kg) for 84 days	→ population levels of adherent and allochthonous bacteria in FG, HG and HC Modulated gut microbiota. ↑ *Psychrobacter*	Ringø et al. [22]a
~24 g	SBM (313 g/kg) for 9 weeks	↑ species richness, Shannon- Weaver index	Dimitroglou et al. [104]
Ctenopharyngodon idella (weight not given)	SBM (13 g/kg) for 8 weeks	↑ *Pseudomonas putida* Aeromonas sp. DH69 *Actinobacterium bacilli* bacterium	Huang [105]
Carassius auratus ♀ × *Cyprinus carpio* ♂ (24.7 g ± 0.4 g)	SBM (300 g/kg) for 8 weeks	→ total culturable aerobic and anaerobic bacteria, presumptive *E. coli*, *Aeromonas*, Bifidobacterium, *Clostridium perfringens*	Cai et al. [106]
Carassius auratus (15 g)	SBM (355 g/kg) for 8 weeks	→ on gut microbiota determined by DGGE	Raggi and Gatlin III [107]
Three cyprinid species	SBM (40 g/kg) for 8 weeks	Modulation of the allochthonous gut microbiota. ↓ Proteobacterium clone (EF707282.1), *Cetobacterium somerae* (AB353124), *Bacillus subtilis*, *Anoxybacillus flavithermus*	Li et al. [108]
Oreochromis niloticus ♀ × *Oreochromis aureus* ♂ (~2 g)	SBM for 8 weeks	↑ *Plesiomonas* sp. BTOK4 *Aeromonas aquarium*	Zhang et al. [109]
Northern snakehead	FM replaced with graded level of SBM and fed for 63 days	At the phylum level, ↓ *Firmicutes* abundance was the lowest in the diet group having 75% defatted fishmeal replacement with SBM, ↑In contrast with *Proteobacteria, Bacteroidetes* and *Planctomycetes* ↓At the genus level, significantly lower abundance of Lactococcus, *Geobacillus*, *Pseudomonas*, *Streptococcus*, *Bacillus* and *Acinetobacter* in diet group (75% defatted fishmeal replacement with SBM) ↑ but higher abundance of *Cetobacterium*, *Planctomyces*, *Shewanella*, *Thermomonas*, *Rubrivivax* and *Carnobacterium* was observed in fish fed the same diet group (75% defatted fishmeal replacement with SBM)	Miao et al. [80]

Table 1. Effects of soy protein inclusion on gut microbiota of fish.

of cultivable intestinal bacteria (aerobic and anerobic). Afterward, by the 8 weeks of feeding trial, the bacterial numbers increased in the FM group, but not in the SBM group. Length heterogeneity analysis of PCR amplified 16S rDNA (LH-PCR) data also suggested a diet-related qualitative change in the intestinal microbiota of fish. The dominant identified genera were among aerobic species *Aeromonas, Sphingomonas,* and *Chryseomonas* and among the lactic acid bacteria, the genera *Lactococcus* and *Lactobacillus*. Rainbow trout fed SBM (450 g/kg) for 16 weeks showed decrease in total culturable species of *Aeromonas* spp., *Vibrio* spp., but the species *Actinomycetales, Psychrobacter* spp., *Saccharomyces* spp. were found as increased number. Total culturable aerobic levels, *Micrococcus* spp., were found unchanged in numbers. Mansfield et al. [96] evaluated the effect of FM and SBM (300 g/kg) on the allochthonous distal intestinal microbiota of triploid female rainbow trout by three cpn60 universal clone libraries, resulting in 1000 and 1181 sequences from FM and SBM, respectively. There were total 32 different sequences were noticed. The most frequently observed sequences were identical to *Carnobacterium (piscicola) maltaromaticum* and accounted for 55 and 97.2% of the clones from the FM and SBM group, respectively. Overall, fish fed FM showed highest diversity (14 different sequences) and only four different sequences observed in the SBM library. In another study, Desai et al. [97] observed that 30% SBM inclusion in rainbow trout diets led to a reduction in *Proteobacteria* and increase in *Firmicutes*. Recently, Bruce et al. [29] evaluated different processed soybean products as a replacement of fishmeal on gut microiota of rainbow trout and observed that the incorporation of processed soy-based proteins alters the microbial community composition within the distal intestine. Species diversity based on abundance and evenness were lowest in the defatted soybean meal group and were significantly less than the bioprocessed soybean meal in low concentration ($p = 0.003$) and commercial soy protein concentrate ($p = 0.003$) treatments.

In Atlantic salmon, fish fed the SBM (250 g/kg) diet had higher total number as well as a more diverse population composition of adherent bacteria in the distal intestine observed by Bakke-McKellep et al. [98]. Green et al. [100] investigated the influence of FM and soybean protein concentrate (SPC; 50 g/kg) on intestinal microbiota of Atlantic salmon. Terminal restriction fragment length polymorphism (T-RFLP) and 16S rRNA clone library analysis revealed that the SPC diet modulated the intestinal microbiome by increasing the bacterial diversity, and a *Pseudomonadales* was more frequently revealed species. In addition, increased *Escherichia coli* also observed in SPC-based diet, but it was absent in FM-based diet. In another study, Navarrete et al. [101] reported SBM supplementation (378 g/kg) effects on distal intestine microbial community of Atlantic salmon. Principal component analysis (PCA) revealed correlations that fish fed SBM diet was correlated with *Aeromonas* VIb and *Sporosarcina aquimarina*, while *Microbacterium, Pseudomonas, Lactococcus lactis sp.* cremoris, and *Aeromonas* VIa were correlated with the FM-based diet. Reveco et al. [103] investigated the microbiota in the mid and distal intestine of Atlantic salmon fed FM and solvent extracted SBM (200 g/kg) by DGGE analysis. Results showed increased *Lactococcus lactis* subsp. lactis in the mid-intestine, while a reduction in *Weissella confusa* in the distal intestine of Atlantic salmon fed 20% solvent extracted SBM-contained diet. Hartviksen et al. [102] revealed no dietary effect of soy protein concentrate (SPC) on total autochthonous bacteria isolated from PI and total allochthonous and total autochthonous bacteria isolated from DI of Atlantic salmon by qPCR analysis. However, significant ($p = 0.05$) effect was observed regarding community composition. An increase was noticed in autochthonous *Enterobacteriaceae*, Bacilli-like, *Lactobacillaceae*, and

Streptococcaceae in PI, and Bacilli-like and *Streptococcaceae* in DI by SPC feeding. In contrast, a significant ($p = 0.05$) decrease was revealed in *Vibrionaceae* in PI.

5.2. In cyprinid fish

Most of the literature available on the effects of different soy products on the gut microbiota are on salmonid fish, and less is known for other species. The possible reasons behind this might be due to the less susceptibility of non-salmonid fish to SBMIE and histological damage [39]. In cyprinid fish, like in grass carp, the effects of dietary SBM inclusion (1.3% by dry weight) were compared with the inclusion of casein meal (CM; 1.0% by dry weight) on the autochthonous gut microbiota [105]. After 8 weeks of feeding, 16S rRNA PCR-DGGE analysis revealed a clear difference between the microbiota of the SBM group and the CM group with similarity between the groups of only 26% ($p < 0.05$). Unique bacteria isolated from the CM group were identified as follows: uncultured *Lachnospiraceae bacterium*, uncultured *Lactobacillus*, uncultured *Clostridium* spp., and uncultured *Proteobacterium*, while bacteria isolated from the SBM group were identified as *Pseudomonas* sp., *Aeromonas* sp., uncultured bacteria, uncultured *Actinobacterium*, and uncultured *Bacillus* spp.

Raggi and Gatlin [107] evaluated four probiotics diets based on FM and SBM on gut microbiota of goldfish (*Carassius auratus*). After 8 weeks of feeding, denaturing gradient gel electrophoresis (DGGE) analysis results revealed no difference in gut microbiota. The probable reason explained for this observation is due to the incorporation of dietary chromic oxide (10 g kg^{-1}) which may have reduced the quantity and complexity of the bacterial community as reported by Ringø [112] for Arctic charr (*Salvelinus alpinus* L.). Cai et al. [106] also reported no significant effects of fishmeal replacement by SBM (30%) on the levels of total aerobic bacteria, total anaerobic bacteria, presumptive *E. coli*, *Aeromonas*, *Bifidobacterium*, or *Clostridium* in the intestine of silver crucian carp (*Carassius auratus gibelio* × *Cyprinus carpio*). Recently, the effect of partial replacement of SBM (4%) by intestinal casing meal (ICM), prepared from the wastewater of enteric coating and heparin processing, was used to evaluate the effect on the allochthonous gut microbiota of three cage-cultured cyprinid species [108]. Results indicated that the allochthonous bacterial diversity was altered by ICM substitution; however, by feeding ICM, some bacterial species were significantly stimulated, *E. coli*, and *Exiguobacterium* in black carp (*Mylopharyngodon piceus*) and species belonging to *Firmicutes*, Fusobacteria, and *Proteobacteria* in gibel carp (*Carassius gebelio*).

5.3. Cichlids and others

The effects of replacing dietary SBM or cottonseed meal (CSM) by completely hydrolyzed feather meal (CHFM) on the composition of gut microbiota was investigated by Zhang et al. [109] for hybrid tilapia. After 8 weeks of feeding, 16S rRNA PCR-DGGE analysis results revealed that CHFM induced modulation of the whole intestinal microbiota in hybrid tilapia and prevented colonization of potentially harmful species in the intestinal tract. *Plesiomonas* sp. BTOK4 and *Aeromonas aquarium* were found in decreased level in diet group where 120 g kg^{-1} CSM was replaced with CHFM. Miao et al. [80] reported the substitution effects of dietary SBM on the intestinal microbial community of northern snakehead. After 63-day feeding, trial results indicated that dietary soybean meal substitutions significantly affected the intestinal microbiota composition of fish. At the phylum level, *Firmicutes* abundance was the lowest in the diet group having 75% defatted fishmeal replacement with SBM, in contrast with *Proteobacteria, Bacteroidetes,* and *Planctomycetes*. At the genus level, significantly lower abundance of *Lactococcus, Geobacillus, Pseudomonas, Streptococcus, Bacillus,* and *Acinetobacter,* but higher abundance of *Cetobacterium, Planctomyces, Shewanella, Thermomonas, Rubrivivax,* and *Carnobacterium* was observed in fish fed the same diet group (75% defatted fishmeal replacement with SBM).

From previous research, it is established that gut microbiota influences several physiological and immunological aspects of aquatic animals like development, digestion, nutrition, immunological functions, and disease resistance [113, 114]. The gut microbiota together with digestive enzymes, mucins, peristalsis, and epithelial barrier with tight junctions belongs to the so-called non-immune component of mucosal immunity [115]. Moreover, several previous research findings indicated that intestinal microbiota is required for full immune maturation [116, 117], inflammatory diseases [117, 118], and to increase the host's resistance toward pathogenic invasion and infection [119]. However, until recently, research relating on the effects of soy protein inclusion in fish feed and their interaction among gut microbiota and immune responses is scarce. So, further research on the interaction effect on gut microbiota and innate immune system due to soy protein utilization are required for further confirmation of the usability of SBM.

6. Conclusion

The highest maximum exploitation of marine resources used to produce FM has enforced fish nutritionist to use alternative protein sources as FM substitute in aquafeed. Worldwide availability and relatively cheaper price make SBM as one of the suitable alternative ingredients in aquafeed. However, high proportion of soy protein sources inclusion in aquafeed may impair fish immunity, maturation, and functionality of the intestinal mucosa, the first line of defense, and damage the gastrointestinal tract. However, using the appropriate proportions of alternative protein sources as well as SBM provides not only the option of limiting harm, but also there is also an interesting possibility to enhance GI immunity and disease resistance. From the available literature, it is showed that non-salmonids are less susceptible on the effects of SBM on the gut microbiota as well as the gut health than salmonids species. Until today, research on the effects of high soy protein inclusion in non-salmonid diets on gut health is little; so, more research warranted for non-salmonids fish. Future study is also needed on the use of different functional supplement in SBM-based diet to increase the efficiency of utilizing alternative protein (soy protein) through maintaining improved physiological and gut health condition. To date, most of the studies on SBM inclusion in aquafeed and its effects on fish intestinal microbiota were descriptive and only concerned the composition of the microbial community. Further works are warranted to investigate the functions of subpopulations in the microbiota and ultimately the functions to the species level due to alternative protein inclusion in aquafeed. In addition, the anaphylactic effects of SBM and the immune regulatory mechanisms involved merits further investigation.

http://dx.doi.org/10.5772/intechopen.92695

Acknowledgements

We gratefully thank Dr. Nicholas Romano (University of Arkansas at Pine Bluff, Pine Bluff, AR, USA) for his help to make figure 3. The authors wish to acknowledge Ajinomoto Co. Inc. (Kanagawa, Japan) for their technical support to make histological slides for Amberjack.

Conflict of interest

The authors declare no conflict of interest.

Author details

Vikas Kumar[1,2*], Md. Sakhawat Hossain[2,3], Janice A. Ragaza[4] and Marina Rubio Benito[1,2]

*Address all correspondence to: vikaskumar@uidaho.edu

1 Department of Animal and Veterinary Science, Aquaculture Research Institute, University of Idaho, Moscow, ID, USA

2 Hagerman Fish Culture Experiment Station, University of Idaho, Hagerman, ID, USA

3 Department of Aquaculture, Faculty of Fisheries, Sylhet Agricultural University, Sylhet, Bangladesh

4 Department of Biology, Ateneo de Manila University, Quezon City, Philippines

References

[1] Kumar V, Bledsoe J, Lee S, Romano N, Small BC, Lalgudi R, et al. Gut Microbiota Homoeostasis Maintains Via Changing the Distal Intestinal Morphology in Rainbow Trout Fed Soy Protein Based Diets. Unpublished; 2020a

[2] Kumar V, Lee S, Cleveland B, Romano N, Lalgudi R, Rubio M, et al. Comparative evaluation of processed soybean meal (EnzoMealTM) vs. regular soybean meal as a fishmeal replacement in diets of rainbow trout (*Oncorhynchus mykiss*): Effects on growth performance and growth-related genes. Aquaculture. 2020b:**516**. DOI: 10.1016/j.aquaculture.2019.734652

[3] Kumar V, Wang H-P, Lalgudi R, Cain R, McGraw B, Rosentrater KA. Processed soybean meal as an alternative protein source for yellow perch (*Perca flavescens*) feed. Aquaculture Nutrition. 2019;**25**(4):917-931

[4] Refstie S, Korsøen ØJ, Storebakken T, Baeverfjord G, Lein I, Roem AJ. Differing nutritional responses to dietary soybean meal in rainbow trout (*Oncorhynchus mykiss*) and Atlantic salmon (*Salmo salar*). Aquaculture. 2000;**190**:49-63

[5] Hossain MS, Koshio S, Ishikawa M, Yokoyama S, Sony MN, Islam MJ, et al. Substitution of dietary fishmeal by soybean meal with inosine administration influences growth, digestibility, immunity, stress resistance and gut morphology of juvenile amberjack *Seriola dumerili*. Aquaculture. 2018;**488**:174-188

[6] Baeverfjord G, Krogdahl Å. Development and regression of soybean meal induced enteritis in Atlantic salmon, *Salmo salar* L., distal intestine: A comparison with the intestines of fasted fish. J. Journal of Fish Diseases. 1996;**19**:375-387

[7] Burrells C, Williams PD, Southgate PJ, Crampton VO. Immunological, physiological and pathological responses of rainbow trout (*Oncorhynchus mykiss*) to increasing dietary concentrations of soybean proteins. Veterinary Immunology and Immunopathology. 1999a;**72**:277-288

[8] Chen W, Ai Q, Mai K, Xu W, Liufu Z, Zhang W, et al. Effects of dietary soybean saponins on feed intake, growth performance, digestibility and intestinal structure in juvenile Japanese flounder (*Paralichthys olivaceus*). Aquaculture. 2011;**318**:95-100

[9] Deng J, Mai K, Ai Q, Zhang W, Wang X, Xu W, et al. Effects of replacing fish meal with soy protein concentrate on feed intake and growth of juvenile Japanese flounder, *Paralichthys olivaceus*. Aquaculture. 2006;**258**:503-513

[10] Kaushik SJ, Cravedi JP, Lalles JP, Sumpter J, Fauconneau B, Laroche M. Partial or total replacement of fish meal by soybean protein on growth, protein utilization, potential estrogenic or antigenic effects, cholesterolemia and flesh quality in rainbow trout, *Oncorhynchus mykiss*. Aquaculture. 1995;**133**:257-274

[11] Krogdahl Å, Bakke-McKellep AM, Baeverfjord G. Effects of graded levels of standard soybean meal on intestinal structure, mucosal enzyme activities, and pancreatic response in Atlantic salmon (*Salmo salar* L.). Aquaculture Nutrition. 2003;**9**:361-371

[12] Li Y, Ai Q, Mai K, Xu W, Cheng Z. Effects of the partial substitution of dietary fish meal by two types of soybean meals on the growth performance of juvenile Japanese seabass, *Lateolabrax japonicus* (Cuvier 1828). Aquaculture Research. 2012;**43**:458-466

[13] Barnes ME, Brown ML, Rosentrater KA, Sewell JR. An initial investigation replacing fish meal with a commercial fermented soybean meal product in the diets of juvenile rainbow trout. Open Journal of Animal Sciences. 2012;**2**:234-243

[14] Chen L, Madl RL, Vadlani PV, Li L, Wang W. Chapter 8: Valueadded products from soybean: Removal of anti-nutritional factors via bioprocessing. In: El-Shemy HA, editor. Soybean Bio-Active Compounds. Rijeka, Croatia: InTechOpen; 2013

[15] Zhou F, Song W, Shao Q, Peng X, Xiao J, Hua Y, et al. Partial replacement of fish meal by fermented soybean meal in diets for black sea bream, *Acanthopagrus schlegelii*, juveniles. Journal of the World Aquaculture Society. 2011;**42**:184-197

[16] Krogdahl Å, Penn M, Thorsen J, Refstie S, Bakke AM. Important antinutrients in plant feedstuffs for aquaculture: An update on recent findings regarding responses in salmonids. Aquaculture Research. 2010;**41**:333-344

[17] Barnes ME, Brown ML, Neiger R. Comparative performance of two rainbow trout strains fed fermented soybean meal. Aquaculture International. 2015;**23**:1227-1238

[18] Kumar V, Makkar HPS, Becker K. Nutritional, physiological and haematological responses in rainbow trout (*Oncorhynchus mykiss*) juveniles fed detoxified Jatropha curcas kernel meal. Aquaculture Nutrition. 2011;**17**:451-467

[19] Kumar V, Makkar HPS, Amselgruber W, Becker K. Physiological, haematological and histopathological responses in common carp (*Cyprinus carpio* L) fingerlings fed with differently detoxified Jatropha curcas kernel meal. Food and Chemical Toxicology. 2010;**48**:2063-2072

[20] Merrifield DL, Olsen RE, Myklebust R, Ringo E. Dietary effect of soybean (*Glycine max*) products on gut histology and microbiota of fish. In: El-Shemy PH, editor. Soybean and Nutrition. Rijeka, Croatia: InTechOpen; 2011. pp. 231-250

[21] Ringø E, Zhou Z, Gonzalez Vecino JL, Wadsworth S, Romero J, Krogdahl A, et al. Effects of dietary components on the gut microbiota of aquatic animals: A never-ending story? Aquaculture Nutrition. 2016a;**22**:219-282

[22] Ringø E, Sperstad S, Myklebust R, Refstie S, Krogdahl A. Characterisation of the microbiota associated with intestine of Atlantic cod (*Gadus morhua* L.). Aquaculture. 2006;**261**: 829-841

[23] Ringø E, Zhou Z, Vecino JLG, Wadsworth S, Romero J, Krogdahl A, et al. Effect of dietary components on the gut microbiota of aquatic animals. A never-ending story? Aquaculture Nutrition. 2016b;**22**:219-282

[24] Cain K, Swan C. Barrier function and immunology. In: Martin Grosell APF, Colin JB, editors. Fish Physiology. Amsterdam, Netherlands: Academic Press; 2010. pp. 111-134

[25] Gómez GD, Balcázar JL. A review on the interactions between gut microbiota and innate immunity of fish. FEMS Immunology and Medical Microbiology. 2008;**52**(2):145-154. DOI: 10.1111/j.1574-695X.2007.00343.x

[26] Liu Y et al. The protective role of glutamine on enteropathy induced by high dose of soybean meal in turbot, *Scophthalmus maximus* L. Aquaculture. 2018;**497**(5):510-519. DOI: 10.1016/j.aquaculture.2018.08.021

[27] Kumar V. Jatropha meal and protein isolate as a protein source in aquafeed [PhD thesis]. Stuttgart, Germany: Department of Aquaculture Systems and Animal Nutrition, for Institute for Animal Productions in the Tropic and Subtropics, University of Hohenheim; 2011. Available from: https://opus.uni-hohenheim.de/volltexte/2011/628/pdf/PhD_thesis_Vikas_Kumar.pdf

[28] Bai N et al. Protective effects of mannan oligosaccharides on turbot Scophthalmus maximus suffering from soy enteropathy. Aquaculture. 2017;**476**(February):141-151. DOI: 10.1016/j.aquaculture.2017.04.005

[29] Bruce TJ, Neiger RD, Brown ML. Gut histology, immunology and the intestinal microbiota of rainbow trout, *Oncorhynchus mykiss* (Walbaum), fed process variants of soybean meal. Aquaculture Research. 2018;**49**:492-504

[30] Gu M et al. Soybean meal induces enteritis in turbot *Scophthalmus maximus* at high supplementation levels. Aquaculture. 2016;**464**:286-295. DOI: 10.1016/j.aquaculture.2016.06.035

[31] Jiang J, Shi D, Zhou X-Q, Hu Y, Feng L, Liu Y. In vitro and in vivo protective effect of arginine against lipopolysaccharide induced inflammatory response in the intestine of juvenile Jian carp (*Cyprinus carpio* var. Jian) fish. Shellfish Immunology. 2015a;**42**(2):457-464

[32] Jiang WD et al. Dietary leucine regulates the intestinal immune status, immune-related signalling molecules and tight junction transcript abundance in grass carp (*Ctenopharyngodon idella*). Aquaculture. 2015b;**444**:134-142. DOI: 10.1016/j.aquaculture.2015.04.005

[33] Zhao S et al. Citric acid mitigates soybean meal induced inflammatory response and tight junction disruption by altering TLR signal transduction in the intestine of turbot, *Scophthalmus maximus* L. Fish & Shellfish Immunology. 2019;**92**(March):181-187. DOI: 10.1016/j.fsi.2019.06.004

[34] Buttle LG, Burrells AC, Good JE, Williams PD, Southgate PJ, Burrells C. The binding of soybean agglutinin (SBA) to the intestinal epithelium of Atlantic salmon *Salmo salar* and rainbow trout, *Oncorhynchus mykiss*, fed high levels of soybean meal. Veterinary Immunology and Immunopathology. 2001;**80**:237-244

[35] Krogdahl A, Bakke-Mckellep AM, Roed KH, Baeverfjord G. Feeding Atlantic salmon *Salmo salar* L. soybean products: Effects on disease resistance (furunculosis), and lysozyme and IgM levels in the intestinal mucosa. Aquaculture Nutrition. 2000a;**6**:77-84

[36] Van den Ingh T, Krogdahl Å, Olli JJ, Hendriks H, Koninkx J. Effects of soybean-containing diets on the proximal and distal intestine in Atlantic salmon (*Salmo salar*): A morphological study. Aquaculture. 1991;**94**:297-305

[37] Van den Ingh TSGAM, Olli JJ, Krogdahl Å. Alcohol-soluble components in soybeans cause morphological changes in the distal intestine of Atlantic salmon, *Salmo salar*, L. Journal of Fish Diseases. 1996;**19**:47-53

[38] Bakke-Mckellep AM, Press CM, Baeverfjord G, Krogdahl Å, Landsverk T. Changes in immune and enzyme histochemical phenotypes of cells in the intestinal mucosa of Atlantic salmon, *Salmo salar* L., with soybean meal-induced enteritis. Journal of Fish Diseases. 2000;**23**:115-127

[39] Zhou Z, Ringø E, Olsen RE, Song SK. Dietary effects of soybean products on gut microbiota and immunity of aquatic animals: A review. Aquaculture Nutrition. 2018;**24**:644-665

[40] Nordrum S, Bakke-McKellep AM, Krogdahl Å, Buddington RK. Effects of soybean meal and salinity on intestinal transport of nutrients in Atlantic salmon (*Salmo salar* L.) and rainbow trout (*Oncorhynchus mykiss*). Comparative Biochemistry and Physiology. B. 2000;**125**:317-335

[41] Urán PA. Etiology of soybean-induced enteritis in fish [PhD thesis]. The Netherlands: Wageningen University; 2008

[42] Booman M, Forster I, Vederas JC, Groman DB, Jones SRM. Soybean meal-induced enteritis in Atlantic salmon (*Salmo salar*) and Chinook salmon (*Oncorhynchus tshawytscha*) but not in pink salmon (*O. gorbuscha*). Aquaculture. 2018;**483**:238-243

[43] Cunningham KE, Turner JR. Myosin light chain kinase: Pulling the strings of epithelial tight junction function. Annals of the New York Academy of Sciences. 2013;**1258**(1): 34-42. DOI: 10.1111/j.1749-6632.2012.06526.x.Myosin

[44] Wang B et al. Glutamine and intestinal barrier function. Amino Acids. 2015;**47**(10):2143-2154. DOI: 10.1007/s00726-014-1773-4

[45] Hossain MS, Koshio S, Ishikawa M, Yokoyama S, Sony NM, Kader MA, et al. Effects of dietary administration of inosine on growth, immune response, oxidative stress and gut morphology of juvenile amberjack, *Seriola dumerili*. Aquaculture. 2017;**468**:534-544

[46] Hossain MS, Koshio S, Ishikawa M, Yokoyama S, Sony NM, Ono S, et al. Comparison of the effects of inosine and inosine monophosphate on growth, immune response, stress resistance and gut morphology of juvenile red sea bream, *Pagrus major*. Aquaculture. 2016;**458**:64-74

[47] Refstie S, Landsverk T, Bakke-McKellep AM, Ringø E, Sundby A, Shearer KD, et al. Digestive capacity, intestinal morphology, and microflora of 1-year and 2-year old Atlantic cod (*Gadus morhua*) fed standard or bioprocessed soybean meal. Aquaculture. 2006;**261**:269-284

[48] Sohrabnezhad M, Sudagar M, Mazandarani M. Effect of dietary soybean meal and multienzyme on intestine histology of beluga sturgeon (*Huso huso*). International Aquatic Research. 2017;**9**:271-280

[49] Wang AR, Ran C, Ringø E, Zhou ZG. Progress in fish gastrointestinal microbiota research. Reviews in Aquaculture. 2018;**10**:626-640

[50] Murashita K, Fukada H, Takahashi N, Hosomi N, Matsunari H, Furuita H, et al. Effect of feed ingredients on digestive enzyme secretion in fish. Bulletin of Japan Fisheries Research and Education Agency. 2015;**40**:69-74

[51] Furutani T, Masumoto T, Fukada H. Response of cholecystokinin and digestive enzyme mRNA levels to various feed ingredients in yellowtail *Seriola quinqueradiata*. Fisheries Science. 2012;**78**:1075-1082

[52] Perera E, Yúfera M. Effects of soybean meal on digestive enzymes activity, expression of inflammation-related genes, and chromatin modifications in marine fish (*Sparus aurata* L.) larvae. Fish Physiology and Biochemistry. 2017;**43**:563-578

[53] Robaina L, Izquierdo MS, Moyano FJ, Socorro J, Vergara JM, Montero D, et al. Soybean and lupin seed meals as protein sources in diets for gilthead seabream (*Sparus aurata*): Nutritional and histological implications. Aquaculture. 1995;**130**:219-233

[54] Santigosa E, Saenz de Rodriganez MA, Rodiles A, Barroso FG, Alarcon FJ. Effect of diets containing a purified soybean trypsin inhibitor on growth performance, digestive

proteases and intestinal histology in juvenile sea bream (*Sparus aurata* L.). Aquaculture Research. 2010;**41**:187-198

[55] Chikwati EM, Sahlmann C, Holm H, Penn MH, Krogdahl Å, Bakke AM. Alterations in digestive enzyme activities during the development of diet-induced enteritis in Atlantic salmon, *Salmo salar* L. Aquaculture. 2013;**402-403**:28-37

[56] Lilleeng E, Froystad MK, Ostby GC, Valen EC, Krogdahl A. Effects of diets containing soybean meal on trypsin mRNA expression and activity in Atlantic salmon (*Salmo salar* L). Comparative Biochemistry and Physiology Part A. 2007;**147**:25-36

[57] Byadgi O, Puteri D, Lee J-W, Chang T-C, Lee YH, Chu CY, et al. The effect of TLR9 agonist CpG oligodeoxynucleotides on the intestinal immune response of cobia (*Rachycentron canadum*). Journal of Immunology Research. 2014;**2014**:273284

[58] Tafalla C, Leal E, Yamaguchi T, Fischer U. T cell immunity in the teleost digestive tract. Developmental and Comparative Immunology. 2016;**64**:167-177

[59] Salinas I, Parra D. Fish mucosal immunity: Intestine. In: Beck B, Peatman E, editors. Mucosal Health in Aquaculture. 1st ed. Amsterdam, Netherlands: Elsevier; 2015. pp. 135-170

[60] Fu GH, Bai ZY, Xia JH, Liu F, Liu P, Yue GH. Analysis of two lysozyme genes and antimicrobial functions of their recombinant proteins in Asian seabass. PLoS One. 2013;**8**:e79743

[61] Sveinbjornsson B, Olsen R, Paulsen S. Immunocytochemical localization of lysozyme in intestinal eosinophilic granule cells of Atlantic salmon, *Salmo* salar L. Journal of Fish Diseases. 1996;**19**:349-355

[62] Shen Y, Zhang J, Xu X, Fu J, Li J. Expression of complement component C7 and involvement in innate immune responses to bacteria in grass carp. Fish & Shellfish Immunology. 2012;**33**:448-454

[63] Kania PW, Sorensen RR, Koch C, Brandt J, Kliem A, Vitved L, et al. Evolutionary conservation of mannan-binding lectin (MBL) in bony fish: Identification, characterization and expression analysis of three bona fide collectin homologues of MBL in the rainbow trout (*Onchorhynchus mykiss*). Fish & Shellfish Immunology. 2010;**29**:910-920

[64] Lovoll M, Kilvik T, Boshra H, Bogwald J, Sunyer JO, Dalmo RA. Maternal transfer of complement components C3-1, C3-3, C3-4, C4, C5, C7, bf, and Df to offspring in rainbow trout (*Oncorhynchus mykiss*). Immunogenetics. 2006;**58**:168-179

[65] Xia JH, Liu P, Liu F, Lin G, Sun F, Tu R, et al. Analysis of stress-responsive transcriptome in the intestine of Asian seabass (*Lates calcarifer*) using RNA-seq. DNA Research. 2013;**20**:449-460

[66] Lokesh J, Fernandes JM, Korsnes K, Bergh O, Brinchmann MF, Kiron V. Transcriptional regulation of cytokines in the intestine of Atlantic cod fed yeast derived mannan oligosaccharide or beta-glucan and challenged with *Vibrio anguillarum*. Fish & Shellfish Immunology. 2012;**33**:626-631

[67] Mulder IE, Wadsworth S, Secombes CJ. Cytokine expression in the intestine of rainbow trout (*Oncorhynchus mykiss*) during infection with *Aeromonas salmonicida*. Fish & Shellfish Immunology. 2007;**23**:747-759

[68] Vasta GR, Nita-Lazar M, Giomarelli B, Ahmed H, Du S, Cammarata M, et al. Structural and functional diversity of the lectin repertoire in teleost fish: Relevance to innate and adaptive immunity. Developmental and Comparative Immunology. 2011;**35**:1388-1399

[69] Casadei E, Bird S, Vecino JL, Wadsworth S, Secombes CJ. The effect of peptidoglycan enriched diets on antimicrobial peptide gene expression in rainbow trout (*Oncorhynchus mykiss*). Fish & Shellfish Immunology. 2013;**34**:529-537

[70] Pan CY, Chen JY, Cheng YS, Chen CY, Ni IH, Sheen JF, et al. Gene expression and localization of the epinecidin-1 antimicrobial peptide in the grouper (*Epinephelus coioides*), and its role in protecting fish against pathogenic infection. DNA and Cell Biology. 2007;**26**:403-413

[71] Seegert D, Rosenstiel P, Pfahler H, Pfefferkorna P, Nikolausa S, Schreibera S. Increased expression of IL-16 in inflammatory bowel disease. Gut. 2001;**48**:326-332

[72] Vojtech LN, Scharping N, Woodson JC, Hansen JD. Roles of inflammatory caspases during processing of zebrafish interleukin-1b in Francisella noatunensis infection. Infection and Immunity. 2012;**80**(8):2878-2885

[73] Larmonier CB, Uno JK, Lee KM, Karrasch T, Laubitz D, Thurston R, et al. Limited effects of dietary curcumin on Th-1 driven colitis in IL-10 deficient mice suggest an IL-10-dependent mechanism of protection. American Journal of Physiology. Gastrointestinal and Liver Physiology. 2008;**295**(5):G1079-G1091

[74] Manzanillo P, Eidenschenk C, Ouyang W. Deciphering the crosstalk among IL-1 and IL-10 family cytokines in intestinal immunity. Trends in Immunology. 2015;**36**(8):471-478

[75] Wang YR, Wang L, Zhang C-X, Song K. Effects of substituting fishmeal with soybean meal on growth performance and intestinal morphology in orange-spotted grouper (*Epinephelus coioides*). Aquaculture Report. 2017;**5**:52-57

[76] Lilleeng E, Penn MH, Haugland O, Xu C, Bakke AM, Krogdahl A, et al. Decreased expression of TGF- beta, GILT and T- cell markers in the early stages of soybean enteropathy in Atlantic salmon (*Salmo salar* L.). Fish & Shellfish Immunology. 2009;**27**:65-72

[77] Sahlmann C, Sutherland BJG, Kortner TM, Koop BF, Krogdahl Å, Bakke AM. Early response of gene expression in the distal intestine of Atlantic salmon (*Salmo salar* L.) during the development of soybean meal induced enteritis. Fish & Shellfish Immunology. 2013;**34**:599-609

[78] Kim DH, Austin B. Innate immune responses in rainbow trout (*Oncorhynchus mykiss*, Walbaum) induced by probiotics. Fish & Shellfish Immunology. 2006;**21**:513-524

[79] Marjara IS, Chikwati EM, Valen EC, Krogdahl Å, Bakke AM. Transcriptional regulation of IL-17A and other inflammatory markers during the development of soybean

meal-induced enteropathy in the distal intestine of Atlantic salmon (*Salmo salar* L.). Cytokine. 2012;**6**:186-196

[80] Miao S, Zhao C, Zhu J, Hu J, Dong X, Sun L. Dietary soybean meal affects intestinal homoeostasis by altering the microbiota, morphology and inflammatory cytokine gene expression in northern snakehead. Scientific Reports. 2018;**8**:113. DOI: 10.1038/s41598-017-18430-7

[81] Gajardo K, Jaramillo-Torres A, Kortner TM, Merrifield DL, Tinsley J, Bakke AM, et al. Alternative protein sources in the diet modulate microbiota and functionality in the distal intestine of Atlantic Salmon (*Salmo salar*) Appl. Environmental Microbiology. 2017;**83**:e02615-e02616. DOI: 10.1128/AEM.02615-16

[82] Austin B. The bacterial microflora of fish, revised. Scientific World Journal. 2006;**6**:931-945

[83] Cahill MM. Bacterial flora of fishes: A review. Microbial Ecology. 1990;**19**:21-41

[84] Hansen GH, Olafsen JA. Bacterial interactions in early life stages of marine cold water fish. Microbial Ecology. 1999;**38**:1-26

[85] Horsley TW. A review of the bacterial flora of teleost and elasmobranchs, including methods for its analysis. Journal of Fish Biology. 1977;**10**:529-553

[86] Izvekova GI, Izvekov EI, Plotnikov AO. Symbiotic microflora in fishes of different ecological groups. The Biological Bulletin. 2007;**34**:610-618

[87] Llewellyn MS, Boutin S, Hoseinifar SH, Derome N. Teleost microbiomes: The state of the art in their characterization, manipulation and importance in aquaculture and fisheries. Frontiers in Microbiology. 2014;**5**:207

[88] Merrifield DL, Dimitroglou A, Foey A, Davies SJ, Baker RR, Bøgwald J, et al. The current status and future focus of probiotic and prebiotic applications for salmonids. Aquaculture. 2010;**302**:1-18

[89] Nya EJ, Austin B. Bacterial microflora of salmonids. In: Polakof S, Moon TW, editors. Trout: From Physiology to Conservation. New York: Nova Science Publishers, Inc.; 2013. pp. 113-129

[90] Ringø E, Birkbeck TH. Intestinal microflora of fish larvae and fry. Aquaculture Research. 1999;**30**:73

[91] Ringø E. Lactic acid bacteria in fish and fish farming. In: Salminen S, Ouwehand A, von Wright A, editors. Lactic Acid Bacteria. New York, NY, USA: Marcel Dekker Inc.; 2004. pp. 581-610

[92] Yoshimizu M, Kimura T. Study on the intestinal microflora of salmonids. Fish Pathology. 1976;**10**:243-259

[93] Kim D-H, Brunt J, Austin B. Microbial diversity of intestinal contents and mucus in rainbow trout (*Oncorhynchus mykiss*). Journal of Applied Microbiology. 2007;**102**:1654-1664

[94] Heikkinen J, Vielma J, Kemilainen O, Tiirola M, Eskelinen P, Kiuru T, et al. Effects of soybean meal based diet on growth performance, gut histopathology and intestinal microbiota of juvenile rainbow trout (*Oncorhynchus mykiss*). Aquaculture. 2006;**261**:259-268

[95] Merrifield DL, Bradley G, Baker RTM, Dimitroglou A, Davies SJ. Probiotic applications for rainbow trout (*Oncorhynchus mykiss* Walbaum) I. Effects on growth performance, feed utilisation, intestinal microbiota and related health criteria. Aquaculture Nutrition. Early View. 2009a;**16**:504-510 DOI: 10.1111/j.1365-2095.2009.00689.x

[96] Mansfield GS, Desai AR, Nilson SA, Van Kessel AG, Drew MD, Hill JE. Characterization of rainbow trout (*Oncorhynchus mykiss*) intestinal microbiota and inflammatory marker gene expression in a recirculating aquaculture system. Aquaculture. 2010;**307**:95-104

[97] Desai AR, Links MG, Collins SA, Mansfield GS, Drew MD, Van Kessel AG, et al. Effects of plant- based diets on the distal gut microbiome of rainbow trout (*Oncorhynchus mykiss*). Aquaculture. 2012;**350-353**:134-142

[98] Bakke-McKellep AM, Penn MH, Salas PM, Refstie S, Sperstad S, Landsverk T, et al. Effects of dietary soybean meal, inulin and oxytetracycline on gastrointestinal histological characteristics, distal intestine cell proliferation and intestinal microbiota in Atlantic salmon (*Salmo salar* L.). British Journal of Nutrition. 2007;**97**:699-713

[99] Ringø E, Sperstad S, Kraugerud OF, Krogdahl A. Use of 16S rRNA gene sequencing analysis to characterize culturable intestinal bacteria in Atlantic salmon (*Salmo salar*) fed diets with cellulose or non-starch polysaccharides from soy. Aquaculture Research. 2008a;**39**:1087-1100

[100] Green TJ, Smullen R, Barnes AC. Dietary soybean protein concentrate- induced intestinal disorder in marine farmed Atlantic salmon, *Salmo salar* is associated with alterations in gut microbiota. Veterinary Microbiology. 2013;**166**:286-292

[101] Navarrete P, Fuentes P, De la Fuentes L, Barros L, Magne F, Opazo R, et al. Short- term effects of dietary soybean meal and lactic acid bacteria on the intestinal morphology and microbiota of Atlantic salmon (*Salmo salar*). Aquaculture Nutrition. 2013;**17**:148-156

[102] Hartviksen M, Gonzalez Vecino JL, Ringø E, Bakke A-M, Wadsworth S, Krogdahl Å, et al. Alternative dietary protein sources for Atlantic salmon (*Salmo salar* L.) affect intestinal microbiota, intestinal and liver histology and growth. Aquaculture Nutrition. 2014;**20**:381-398

[103] Reveco FE, Øverland M, Romarheim OH, Mydland LT. Intestinal bacterial community structure differs between healthy and inflamed intestines in Atlantic salmon (*Salmo salar* L.). Aquaculture. 2014;**420-421**:262-269

[104] Dimitroglou A, Merrifield DL, Spring P, Sweetman J, Moate R, Davies SJ. Effects of mannan oligosaccharides (MOS) supplementation on growth performance, feed utilisation, intestinal histology and gut microbiota of gilthead sea bream (*Sparus aurata*). Aquaculture. 2010a;**300**:182-188

[105] Huang G. The Study of Intestinal Bacterial Molecular Ecology of Cultured Fishes [master thesis]. Huazhong Agricultural University (in Chinese); 2008

[106] Cai C-F, Wang W-J, Ye Y-T, Krogdahl Å, Wang Y-L, Xia Y-M, et al. Effect of soybean meal, raffinose and stachyose on the growth, body composition, intestinal morphology and intestinal microflora of juvenile allogynogenetic silver crucian carp (*Carassius auratus* ♀ x *Cyprinus carpio* ♂). Aquaculture Research. 2012;**43**:128-138

[107] Raggi T, Gatlin DM III. Prebiotics have limited effects on nutrient digestibility of a diet based on fish meal and soybean meal in goldfish. North American Journal of Aquaculture. 2012;**74**:400-407

[108] Li Z, Xu L, Liu W, Liu Y, Ringø E, Du Z, et al. Protein replacement in practical diets altered gut allochthonous bacteria of cultured cyprinid species with different food habits. Aquaculture International. 2015;**23**:913-928

[109] Zhang Z, Xu L, Liu W, Yang Y, Du Z, Zhou Z. Effects of partial replacing soybean meal or cottonseed meal with completely hydrolysed feather meal (defatted rice bran as the carrier) on production, cytokines, adhesive gut bacteria, and disease resistance in hybrid tilapia (*Oreochromis niloticus* ♀ x *Oreochromis aureus* ♂). Fish & Shellfish Immunology. 2014;**41**:517-525

[110] Tamboli CP, Neut C, Desreumaux P, Colombel JF. Dysbiosis as a prerequisite for IBD. Gut. 2004;**53**:1057

[111] Turroni F, Ventura M, Buttó LF, Duranti S, O'Toole RW, Motherway MO, et al. Molecular dialogue between the human gut microbiota and the host: A *Lactobacillus* and *Bifidobacterium* perspective. Cellular and Molecular Life Sciences. 2014;**71**:183-203

[112] Ringø E. Arctic charr, *Salvelinus alpinus* (L.), reared in fresh and sea water. An experimental study of lipid digestion and intestinal microflora [Ph.D. thesis]. Norway: Norwegian College of Fishery Science, University of Tromsø; 1993

[113] Montalban-Arques A, De Schryver P, Bossier P, Gorkiewicz G, Mulero V, Gatlin DM III, et al. Selective manipulation of the gut microbiota improves immune status in vertebrates. Frontiers in Immunology. 2015;**6**:512. DOI: 10.3389/fimmu.2015.00512

[114] Romero J, Ringø E, Merrifield DL. The gut microbiota of fish. In: Merrifield D, Ringø E, editors. Aquaculture Nutrition: Gut Health, Probiotics and Prebiotics. Oxford, UK: Wiley-Blackwell Publishing; 2014. pp. 75-100

[115] Purchiaroni F, Tortora A, Gabrielli M, Bertucci F, Gigante G, Ianiro G, et al. The role of intestinal microbiota and the immune system. European Review for Medical and Pharmacological Sciences. 2013;**17**:323-333

[116] Chung H, Pamp SJ, Hill J, Surana NK, Edelman SM, Troy EB, et al. Gut immune maturation depends on colonization with a host- specific microbiota. Cell. 2012;**149**:1578-1593

[117] Kamada N, Seo S-U, Chen GY, Núnez G. Role of the gut microbiota in immunity and inflammatory disease. Nature Reviews Immunology. 2013;**13**:321-335

[118] Rendueles O, Ferrieres L, Fretaud M, Begaud E, Herbomel P, Levraud J-P, et al. A new zebrafish model of oro- intestinal pathogen colonization reveals a key role for adhesion in protection by probiotic bacteria. PLoS Pathogens. 2012;**8**:7. DOI: 10.1371/journal.ppat.1002815

[119] De Schryver P, Vadstein O. Ecological theory as a foundation to control pathogenic invasion in aquaculture. ISME Journal. 2014;**8**:2360-2368

Chapter 6

Nondestructive Evaluation of Inoculation Effects of AMF and *Bradyrhizobium japonicum* on Soybean under Drought Stress from Reflectance Spectroscopy

Weiping Kong, Yinli Bi, Wenjiang Huang,
Lingli Tang, Chuanrong Li and Lingling Ma

Additional information is available at the end of the chapter

http://dx.doi.org/10.5772/intechopen.88673

Abstract

Precise estimation of leaf chlorophyll content (LCC) and leaf water content (LWC) of soybean, using remote sensing technology, provides a new avenue for the nondestructive evaluation of inoculation effects of arbuscular mycorrhizal fungi (AMF) and *Bradyrhizobium japonicum* (BJ) on soybean growth condition. In this study, a series of pot experiments were conducted in the greenhouse, soybean inoculated with *Glomus intraradices* (G.i, one of AMF species), G.i and BJ, and non-inoculation were planted under drought stress (DS) and normal irrigation (NI) conditions. Leaf spectra and LCC and LWC were measured on the 28th and 56th days after inoculation. Two new simple ratio (SR) indices, derived from the first derivative spectral reflectance at λ1 nm ($D_{\lambda 1}$) and the raw spectral reflectance at λ2 nm ($R_{\lambda 2}$), were developed to estimate LCC and LWC. The results indicate that under DS, plants inoculated with G.i had higher LCC and LWC than the non-inoculated plants, followed by the counterparts co-inoculated with G.i and BJ. Linear estimation models, established by the $D_{650}/R_{red\ edge}$ and D_{1680}/R_{680}, achieved great improved accuracy for quantifying LCC and LWC of soybean under inoculation and drought stress treatments, with determination of coefficient of 0.63 and 0.76, respectively.

Keywords: leaf chlorophyll, leaf water, remote sensing, soybean, inoculation effects

1. Introduction

Drought is one of the major abiotic stress factors, which reduces plant growth, productivity, and other physiological processes of soybean around the world. For the leguminous plants, there are a lot of AMF and rhizobium in their rhizosphere micro-ecosystem [1]. More than 80% of land plants' roots are known to establish a mutualistic association with AMF [2, 3]. An increasing number of studies have indicated that the symbiosis is beneficial to both the host plants and AMF. The host plants provide the fungi directly with needed carbohydrates including glucose and sucrose to maintain their life cycle [4]. In turn, the fungi can expand the absorption range and area of plant roots and improve plant uptake of water and nutrients in soil [5], greatly contributing to enhance the resistance of plants to drought stress [6].

Over recent years, substantial efforts have focused on understanding the mechanisms of inoculation effects on moisture and nutrient uptake of host plants and the enhanced stress tolerance (e.g., drought stress). Numerous studies have shown the positive effects of the inoculation on plant growth. The symbiotic effects of AMF and BJ on leguminous plants were tested; studies reported that AMF and rhizobium were both beneficial and constrained to each other [7, 8]. The results of Auge et al. [6] and Abdel-Salam et al. [9] provide evidences that AMF improved drought tolerance of rose plants. Many researchers have found that a series of leaf parameters including chlorophyll a and chlorophyll b contents, LCC, LWC, and leaf area of AMF inoculated plants had increased, compared to non-inoculated plants under abiotic stresses [10–12]. Given the sensitivity of LCC and LWC to the inoculation, they are considered as the main biochemical parameters to indicate the inoculation effects of AMF and BJ on plant growth. However, these indicators are almost measured with traditional chemical methods in the laboratory, which is time-consuming and destructive sampled. An accurate, rapid, and nondestructive method is urgently needed for the direct or indirect evaluation of inoculation effects.

Because leaf spectral reflectance is predominantly influenced by LCC and leaf cell structures in visible to near-infrared (NIR) regions (400–900 nm) and by LWC in NIR to shortwave infrared (SWIR) regions (900–2500 nm), the estimation of LCC and LWC with nondestructive optical method, i.e., remote sensing technology, has been developed, suggesting that it would be possible to offer a new avenue to indirectly monitor the effects of AMF and BJ inoculation. Several approaches have been applied to assess LCC and LWC of plants from their optical properties, such as spectral indices or spectral transformations. Chappelle et al. [13] suggested that combinations of band ratios can minimize the effects of spectral convolution; lots of simple ratio (SR) type of spectral indices were proposed in the literatures. Indices, such as pigment-specific simple ratio (PSSR), red-edge spectral parameter (RES), water index (WI), reciprocal of moisture stress index (RMSI), moisture stress index (MSI), etc., were sensitive to LCC and LWC of plants and widely used for the estimation [14–18]. In addition, the first derivative transformation of the apparent absorbance was found to show a great potential for leaf biochemical parameter monitoring [19]. Several derivative spectral indices were derived from the first derivative spectra [20, 21].

So far, only limited information is available on remote estimation of LCC and LWC of soybean inoculated with AMF and/or BJ and on nondestructive and indirect evaluation of the inoculation effects of AMF and/or BJ using remote sensing data. The objectives of this study were (i) to investigate the effects of inoculation on LCC and LWC of soybean under the inoculation and drought stress treatments; (ii) to identify the optimum bands and develop spectral indices for the precise estimation of LCC and LWC, from the first derivative spectra and the raw spectra of soybean leaves; and (iii) to establish the estimation models for LCC and LWC in leaves with AMF inoculation and AMF and BJ co-inoculation, respectively.

2. Materials and methods

2.1. Experimental design

The experiment was conducted in the microbial remediation greenhouse at China University of Mining and Technology (Beijing) (116°21.3′ E, 40°00′ N), from the middle of May to the beginning of July, in 2014. A cultivar of soybean seeds (*Zhonghuang* 35), provided by the Chinese Academy of Agricultural Sciences, were selected and sowed in the pots. The tested strains were arbuscular mycorrhizal *Glomus intraradices* and *Bradyrhizobium japonicum*. 1.1 kg of sandy soil was filled in per pot after sieving, autoclaving, and air-drying. The pot size was 12 cm (top diameter) × 9 cm (bottom diameter) × 15 cm (height). The basic characteristics of sandy soil were as follows: 7.62 of pH, 35.4 μs cm^{-1} of electrical conductivity, 26% of maximum water holding capacity, 13.75 mg kg^{-1} of available phosphorus, and 49.23 mg kg^{-1} of available potassium. Before sowing, the nutrient solution containing NH_4NO_3, KH_2PO_4, and KNO_3 was applied prior to seeding as basal fertilizer, which made the mass fraction of nitrogen, phosphorus, and potassium in the soil reach to 100×10^{-6}, 10×10^{-6}, and 150×10^{-6}, respectively. After 3 weeks of emergence, two healthy and uniformed soybean seedlings were retained per pot, half of the pots were subjected to drought stress (DS), and the other half were irrigated normally (NI), with the maximum water holding capacity of 35 and 75%, respectively. There were G.i inoculation, G.i and BJ inoculation (G.i + BJ), and non-inoculation (CK) treatments under each water gradient. For two inoculation treatments, 50 g of G.i strains soil was added to each pot, while 50 g autoclaved sandy soil was added to the pots of CK treatment. Additionally, for the G.i + BJ treatment, 10 ml BJ solution was poured around the plant roots on the first day of emergence. Each treatment had four replications and was harvested two times, on the 28th and 56th days of inoculation, respectively. All treatments were randomly arranged in the greenhouse.

2.2. Data acquisition

Leaf spectral reflectance and leaf biochemical parameters were measured on the 28th day and 56th day after inoculating, the 28th day of inoculation was exactly the 7th day after drought stress. Forty-five datasets were obtained after removing the outliers.

2.2.1. *Leaf spectral reflectance measurement*

All leaf spectral reflectance of soybean measurements was made with an ASD FieldSpec4 spectrometer (Analytical Spectral Devices, Boulder, CO, USA) in a dark room. The instrument was fitted with a 25° field of view fiber optics, recording reflectance between 350 nm and 1050 nm with a sampling interval of 1.40 nm and a resolution of 3 nm and reflectance between

1000 and 2500 nm with a sampling interval of 2 nm and a resolution of 10 nm. For each pot, two leaves were cut from the top, middle, and bottom layers of the plants, respectively. Spectral measurement was made at the height of 20 cm above each leaf from four angles rotated with a step of 90°. Each spectral measurement was preceded by an optimization measurement, and a white reference measurement was taken before leaf spectral measurement using a white Spectralon® (Labsphere, Inc., New Hampshire, USA) reference panel. Ten scans were performed for each angle, then 40 scans were averaged as the spectral reflectance of each leaf. The averaged reflectance of six leaves was used to represent the plants' spectrum of each pot.

2.2.2. *Leaf biochemical parameter measurement*

Immediately following spectral reflectance measurement, LCC and leaf fresh weight (LFW) were measured. LCC was represented by values obtained by SPAD-502 (Minolta Camera Co., Osaka, Japan), due to the good correlation with the extractable chlorophyll content [22]. Fifteen to Twenty points were selected randomly on the leaf; their average value was considered as the LCC.

LFW and leaf dry weight (LDW) were measured using the analytical balance. After weighting LFW, leaves were dried at 105°C for 30 min in an oven and subsequently dried at 80°C until constant weight to measure LDW. LWC (%) was calculated as.

$$\mathrm{LWC}\,(\%) = (FW - DW)/FW \times 100\% \tag{1}$$

2.3. Spectral first derivative transformation

Spectral first derivative transformation was used to process the raw spectral reflectance from 400 nm to 2500 nm. This was computed by Eq. (2), allowing for suppression of the effects from illumination, background reflectance, atmospheric scattering, and absorption on leaf spectral reflectance and inversely highlighting spectral absorption features of leaf biochemical parameters [23]:

$$\mathrm{D}_{\lambda} = (\mathrm{R}_{\lambda-1} + \mathrm{R}_{\lambda+1})/2\Delta\lambda \tag{2}$$

where D_λ is the first derivative spectral reflectance at wavelength λ, $R_{\lambda-1}$ and $R_{\lambda+1}$ are the raw spectral reflectance before and after wavelength λ, and $\Delta\lambda$ is the interval between wavelength $\lambda - 1$ and wavelength $\lambda + 1$.

2.4. The construction of spectral indices

SR is one of the primary types of indices for vegetation biochemical parameters. An SR index is conventionally defined as a ratio of the raw spectral reflectance at two wavelengths ($R_{\lambda1}$ and $R_{\lambda2}$); its formulation is as Eq. (3). The derivative reflectance has shown to be more sensitive to leaf biochemical parameters at corresponding absorption wavebands than the raw spectral reflectance [19]. Inspired, we inserted a first derivative spectral reflectance ($D_{\lambda1}$) into SR and constructed the new SR indices with a raw reflectance ($R_{\lambda 2}$) to quantify LCC and LWC for the inoculated soybean, shown as Eq. (4).

$$SR = R_{\lambda 1}/R_{\lambda 2} \quad (3)$$

$$SR_{D_\lambda} = D_{\lambda 1}/R_{\lambda 2} \quad (4)$$

2.5. Statistical analysis

Correlation analysis was used to analyze the relationship between reflectance and LCC and LWC of soybean. Linear regression was applied to model the relationship between spectral indices and LCC and LWC. A leave-one-out cross-validation procedure was used to validate the behavior of spectral indices in LCC and LWC estimation, due to the limited number of sample datasets in our case. The coefficient of determination (R^2), root mean square error (RMSE), and mean absolute error (MAE) were employed as accuracy indicators to evaluate the performances of estimation models, as well as the accuracy of reflection of inoculation effects. The higher the R^2 and the lower the RMSE and MAE, the better accuracy of the models. RMSE and MAE were computed as follows:

$$RMSE = \sqrt{\sum_{i=1}^{n} (\hat{y}_i - y_i)^2 / n} \quad (5)$$

$$MAE = \frac{1}{n} \sum_{i=1}^{n} \left| \hat{y}_i - y_i \right| \quad (6)$$

3. Results and discussion

3.1. The effects of inoculation on leaf chlorophyll content and leaf water content

The changes of LCC of soybean under different treatments at two inoculation periods are shown in **Table 1**. LCC of inoculated soybean under NI (i.e., G.i + BJ-NI and G.i-NI) increased, whereas LCC under all other treatments decreased as time went on. After 28 days of inoculation, comparison of LCC of G.i + BJ-DS and G.i-DS with lower content of CK-DS revealed a significant difference due to the effects of inoculation. However, under NI treatment, the difference of LCC between inoculation and CK had not reached the significant level. After 56 days, drought stress resulted in a reduced LCC of soybean, since the stressed plants all had lower LCC than the corresponding normal irrigated plants. LCC of soybean co-inoculated with G.i + BJ and G.i was further significantly higher than that CK under NI, reaching as high as 43.63 and 43.23, respectively.

LWC of soybean under different treatments was measured at two inoculation periods; they were highly affected by the inoculation and application of drought stress (**Figure 1**). After 28 days of inoculation, the LWC was relatively high, with the maximum of 82.3% under CK-NI treatment. Similar to LCC, drought also led to a strong decline in LWC during the beginning of stress. This was reflected by the lower LWC of three treatments under DS compared to the corresponding treatments under NI. After 56 days of inoculation, LWC of all treatments sharply declined along the plant growth, especially for the G.i + BJ-NI and G.i-NI treatments (more than 11%). The growth process of plants markedly increased in July, with the improvement of nutritional and moisture uptake and transport, the growth of more leaves, and bigger leaf area. The stomatal opening and plant transpiration became extremely enhanced, due to the bigger leaf area as well as considerable high temperature in the greenhouse caused by the high outdoor temperature and the plant respiration, consequently resulting in the reduction of LWC. Such decrease in LWC was more pronounced in G.i + BJ/G.i plants than CK plants under NI, partly because the former had much bigger leaf area and more transpiration than the latter.

Treatments	28 days	56 days
G.i + BJ-DS	39.88a	36.73b
G.i-DS	38.28a	36.37b
CK-DS	34.26b	34.15b
G.i + BJ-NI	37.63a	43.63a
G.i-NI	38.79a	43.23a
CK-NI	38.41a	36.27b

Note: G.i + BJ, G.i, and CK indicate treatments of G.i and BJ co-inoculation, G.i inoculation, and non-inoculation, respectively; DS and NI indicate drought stress and normal irrigation, respectively. G.i + BJ-DS indicates soybean co-inoculated with G.i and BJ under drought stress, etc. The letters (a and b) indicate significant differences at $p < 0.05$ among six treatments at each inoculation period.

Table 1. Soybean leaf chlorophyll content under different treatments at two inoculation periods.

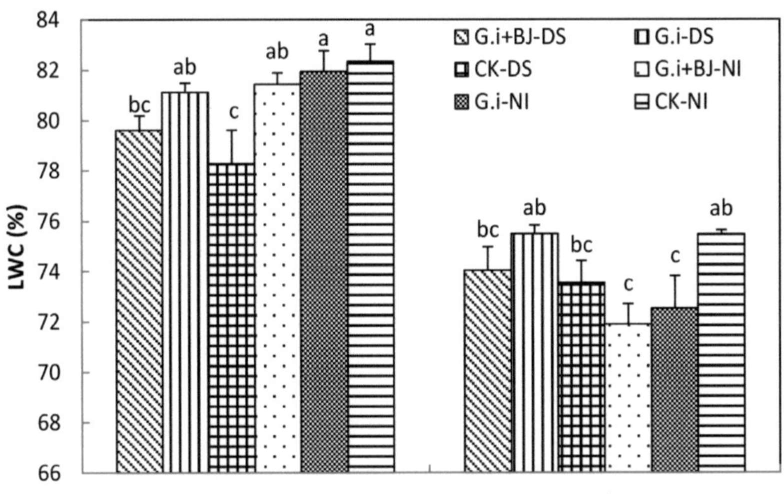

Figure 1. Soybean leaf water content (LWC) under different treatments at two inoculation periods. G.i + BJ, G.i, and CK indicate treatments of G.i and BJ co-inoculation, G.i inoculation, and non-inoculation, respectively; DS and NI indicate drought stress and normal irrigation, respectively. G.i + BJ-DS indicates soybean co-inoculated with G.i and BJ under drought stress, etc. The letters (a–c) indicate significant differences at $p < 0.05$ among six treatments at each inoculation period.

For the LCC and LWC, it was worth noting that the G.i and G.i + BJ plants achieved higher LCC and LWC than CK plants under DS during the two inoculation periods, which was in accordance with the finding of Aliasgharzad et al. [24]. This often benefits from the a mutualistic association formed by AMF and soybean roots (or AMF, soybean roots, and BJ), which could help plants to alleviate the suffering of drought stress directly [9, 25]. Another reason for enhancing drought tolerance in the inoculated plants might be the improvement of phosphorus absorption [26]. In comparison to G.i treatment, LCC and LWC of G.i + BJ plants were slightly lower, because the relationship of mutual restriction between G.i and BJ might play a dominant role compared to that of mutual promotion [7].

3.2. Identifying optimum bands for spectral indices

We analyzed the correlation between first derivative reflectance and LCC and LWC from 400 to 2500 nm, respectively. The results are presented in **Figures 2a** and **3a**. Generally, the derivative of the green to NIR region (around 550–800 nm) appeared to correlate well with LCC (**Figure 2a**), whereas LWC showed relative high correlation with the derivative reflectance in the NIR to shortwave region (around 1400–2000 nm) compared to the visible region. Specifically, it was evident that LCC got higher coefficients ($-0.5 < r < 0.5$) at three spectral regions, i.e., 610–620, 645–665, and 730–760 nm. LWC showed high r values ($-0.5 < r < 0.5$) with derivative spectra of 1410–1420, 1675–1685, 1875–1880, 1900–1910, and 2240–2280 nm, since they are the absorption bands of leaf water. It was interesting to note that LWC also obtained good correlation in the region between 720 and 750 nm, probably due to its covariance with LCC.

The red-edge position (REP) is defined as the wavelength which has the maximum reflectance of the derivative spectrum within the range of 680–760 nm [27]; it is the transition between low reflectance in the red region and high reflectance in the NIR of the raw spectrum [28]. The reflectance of red-edge region was found to be frequently impacted by chlorophyll absorption [29], affecting many spectral indices (e.g., SR). We, therefore, chose the raw spectral reflectance of REP (i.e., R_{REP}) as the denominator of SR and $SR_{D\lambda}$ indices for soybean LCC estimation. To identify the optimal band of $D_{\lambda 1}$ in the $SR_{D\lambda}$ index, we calculate linear regression of the $SR_{D\lambda}$ ($D_{\lambda 1}/R_{REP}$) model vs. LCC, with λ1 changed from 610 to 760 nm; the result is shown in **Figure 2b**. It was pronounced in LCC estimation where bands between 640 and 660 nm show the higher coefficients of determination (R^2), with peak at 650 nm, near the band of leaf chlorophyll maximum absorption. Then the $SR_{D\lambda}$ index composed of D_{650} and R_{REP} held a promising potential for LCC assessment.

Previous studies have indicated that the significant spectral indices for LWC estimation tend to link to LCC, because of the covariation between the two leaf biochemical parameters [30]. Our result showed the similarity with such finding, with correlation coefficient of −0.4 (not shown for brevity). As a result, 680 nm, where the maximum absorption of chlorophyll, was selected, and R_{680} was used in $SR_{D\lambda}$ index to estimate LWC. Similar to LCC, the R^2 of $SR_{D\lambda}$ ($D_{\lambda 1}/R_{680}$) linear models vs. LWC was computed, with λ1 changed from 1410 to 2280 nm. As shown in **Figure 3b**, three peaks with higher R^2 values were achieved where λ1 was centered around 1410, 1680, and 1900 nm. It was evident that the $SR_{D\lambda}$ index employed the derivative reflectance of 1680 nm (D_{1680}) as the numerator was the most suitable for LWC estimation.

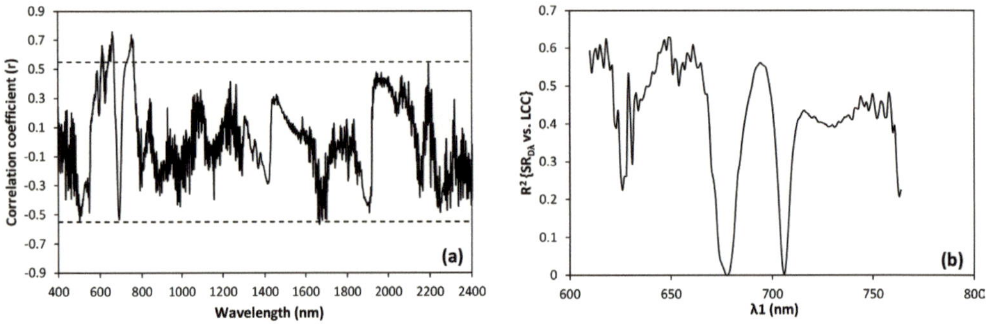

Figure 2. (a) Correlation coefficient (r) between first derivative reflectance and LCC, the dotted lines indicate r = 0.55 or r = −0.55; (b) the R^2 of linear estimation models between $SR_{D\lambda}$ and LCC with λ1 in $D_{\lambda 1}$ changed from 610 to 760 nm.

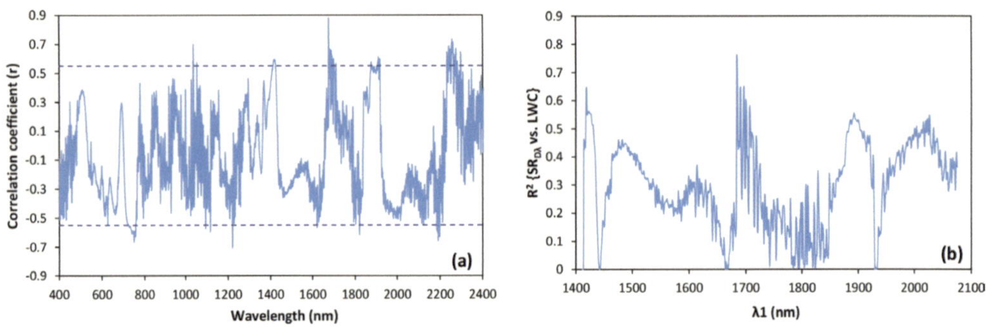

Figure 3. (a) Correlation coefficient (r) between first derivative reflectance and LWC; the dotted lines indicate r = 0.55 or r = −0.55; (b) the R^2 of linear estimation models between $SR_{D\lambda}$ and LWC with λ1 in $D_{\lambda 1}$ changed from 1410 to 2280 nm.

3.3. Leaf chlorophyll content estimation and validation

To indirectly and nondestructively evaluate the inoculation effects on the growth of soybean under all treatments, we estimated LCC and LWC using remote sensing data. Empirical relationships between spectral indices and LCC and LWC were proposed. Based on the analyses presented above, the SR and $SR_{D\lambda}$ indices, developed in this study, for LCC and LWC estimation, were formulated as Eqs. (7)–(10); they were referred as SR-LCC, $SR_{D\lambda}$-LCC, SR-LWC, and $SR_{D\lambda}$-LWC, respectively.

$$\text{SR-LCC} = R_{650}/R_{REP} \tag{7}$$

$$SR_{D_\lambda}\text{-LCC} = D_{650}/R_{REP} \tag{8}$$

$$\text{SR-LWC} = R_{1680}/R_{680} \tag{9}$$

$$SR_{D_\lambda}\text{-LWC} = D_{1680}/R_{680} \tag{10}$$

Forty-five datasets were used to build the estimation models of LCC under different treatments; the results are shown in **Figure 4**. From the distribution of scattering points of the two models, the new index $SR_{D\lambda}$-LCC was strongly and linearly related to LCC variation, whereas the SR-LCC was less sensitive to LCC with the points scattering around the fitted line.

On the other hand, the $SR_{D\lambda}$-LCC model greatly outperformed the SR-LCC model to assess the changes of LCC, achieving the higher value of the coefficient of determination ($R^2 = 0.63$), compared to the R^2 value of 0.11. Furthermore, these results provide a definite proof that spectral first derivative transformation plays an important role in LCC estimation.

To evaluate the general applicability of spectral indices and the prediction precision of models, we analyzed the validation results of LCC (**Figure 5**). As expected from the model calibration, the SR-LCC yielded the weaker prediction accuracy than the $SR_{D\lambda}$-LCC, with a lower R^2 and higher RMSE and MAE values (R^2, RMSE, and MAE of 0.02, 4.25, and 3.58%). Nevertheless, the $SR_{D\lambda}$-LCC model was superior to the SR-LCC model, with the R^2 of 0.59 and RMSE and MAE of 2.68, and 2.15%, demonstrating that it was the best model for the estimation of LCC inoculated with BJ and/or G.i.

3.4. Leaf water content estimation and validation

The linear relationship between LWC of soybean and spectral indices were plotted in **Figure 6**. As seen, the $SR_{D\lambda}$-LWC had a significant linear relationship with LWC; 76% of variation was

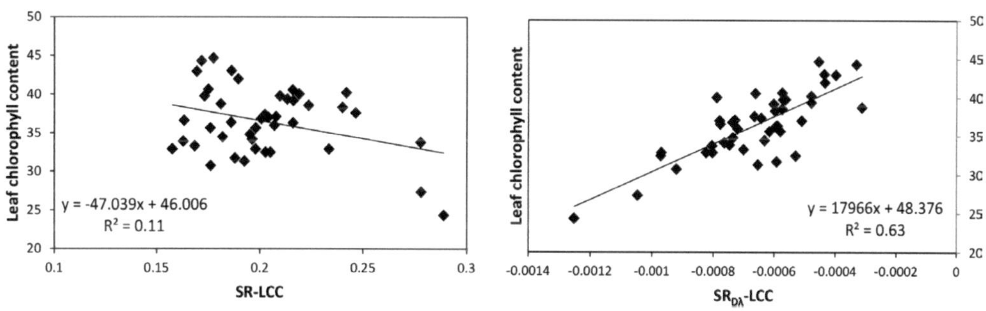

Figure 4. Relationship between leaf chlorophyll content and spectral indices under inoculation and drought stress treatments.

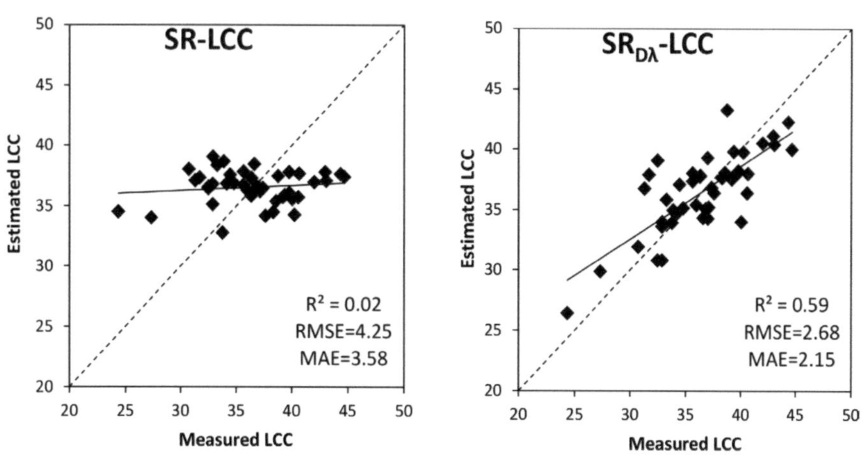

Figure 5. Relationship between measured LCC and estimated LCC using the SR-LCC and $SR_{D\lambda}$-LCC indices under inoculation and drought stress treatments.

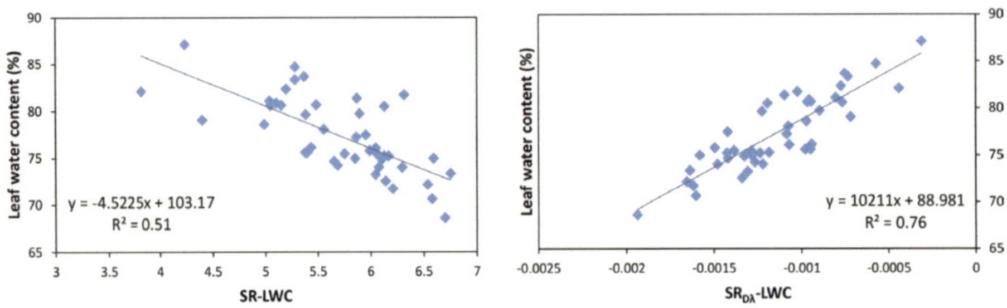

Figure 6. Relationship between leaf water content and spectral indices under inoculation and drought stress treatments.

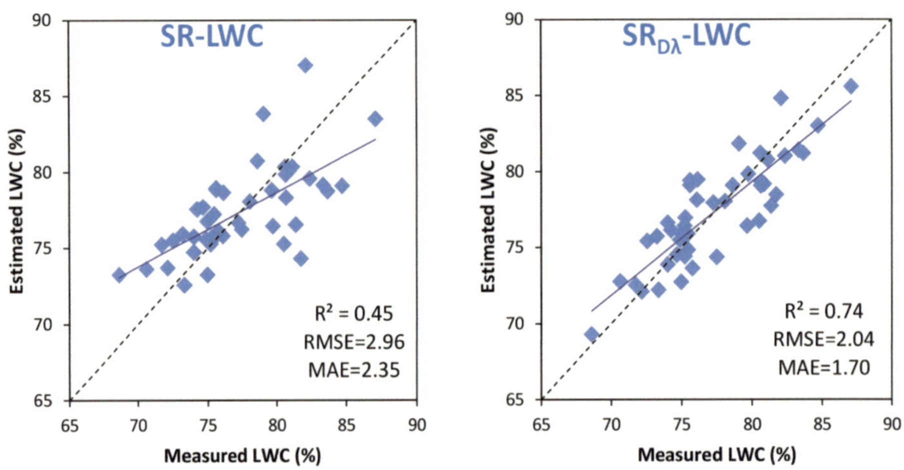

Figure 7. Relationship between measured LWC and estimated LWC using the SR-LWC and $SR_{D\lambda}$-LWC indices under inoculation and drought stress treatments.

explained by the estimation model. Compared to the SR-LWC, the $SR_{D\lambda}$-LWC index yielded a significant increase in R^2 by 49%, greatly improving the accuracy of estimation of LWC under inoculation and drought stress treatments. Meanwhile, it should be noted that the SR index performed better in modeling LWC than its behavior in estimating LCC (**Figure 4**).

Also in this case, the predictive power of spectral indices for LWC estimation were tested (**Figure 7**). The scattering plot showed that a consistent agreement between LWC measured in the laboratory and those estimated by $SR_{D\lambda}$-LWC index, with significant R^2, RMSE, and MAE values of 0.74, 2.04, and 1.70%, respectively, showing its strong stability and predictability. The result indicated that the $SR_{D\lambda}$-LWC model was reliable to predict LWC in this study.

4. Conclusion

The analyses of this study indicated the potential of remote sensing technology in the evaluation of inoculation effects of arbuscular mycorrhizal fungi (AMF) and *Bradyrhizobium japonicum* on soybean growth condition. Drought reduces nutrient uptake, plant growth, and other physiological processes of most crops. Given the higher leaf chlorophyll content (LCC) and leaf water content (LWC) in G.i and BJ-co-inoculated and G.i-inoculated plants than non-inoculated plants under drought stress after 28 and 56 days of inoculation in this study, it can be concluded that rhizobium and/or mycorrhizal fungi could contribute to increase drought tolerance ability of plants, thus leading to alleviation of drought stress in soybean. The changes of LCC and LWC could reflect the inoculation effects on soybean growth condition. Spectral first derivative transformation produced more effective spectral indices for LCC and LWC assessment. The two $SR_{D\lambda}$ indices, using the reflectance of first derivative spectra and the raw spectra of leaves, have been proposed and applied. We found that the $SR_{D\lambda}$-LCC (D_{650}/R_{REP}) and $SR_{D\lambda}$-LWC (D_{1680}/R_{680}) were highly sensitive to LCC and LWC, as compared to the indices that have analogous forms using two reflectances corresponding to the same wavebands of raw spectra, showing their advantages over other spectral indices. Accordingly, quantifying LCC and LWC of crops using remote sensing data would offer a nondestructive and indirect way to monitor the inoculation effects; meanwhile, our results could also expand the present capabilities and applications of vegetation remote sensing.

Acknowledgements

This work was supported by the National Key Research and Development Program of China (2016YFB0501501) and the Strategic Priority Research Program of the Chinese Academy of Sciences (XDA13030402). The authors are grateful to the reviewers for their helpful comments. The authors also would like to thank all the colleagues that contributed to the experiment.

Conflict of interest

The authors declare no conflict of interest.

Author details

Weiping Kong[1]*, Yinli Bi[2], Wenjiang Huang[3], Lingli Tang[1], Chuanrong Li[1] and Lingling Ma[1]

*Address all correspondence to: kongwp@radi.ac.cn

1 Key Laboratory of Quantitative Remote Sensing Information Technology, Aerospace Information Research Institute, Chinese Academy of Sciences, Beijing, China

2 State Key Laboratory of Coal Resources and Safe Mining, China University of Mining and Technology (Beijing), Beijing, China

3 Key Laboratory of Digital Earth Science, Aerospace Information Research Institute, Chinese Academy of Sciences, Beijing, China

References

[1] Antunes PM, de Varennes A, Rajcan I, Goss MJ. Accumulation of specific flavonoids in soybean (*glycine max* (L.) *merr.*) as a function of the early tripartite symbiosis with arbuscular mycorrhizal fungi and bradyrhizobium japonicum (kirchner) Jordan. Soil Biology and Biochemistry. 2006;**38**:1234-1242

[2] Smith SE, Read D. Mycorrhizal Symbiosis. Cambridge: Academic Press; 2008

[3] Pozo MJ, Azcon-Aguilar C. Unraveling mycorrhiza-induced resistance. Current Opinion in Plant Biology. 2007;**10**:393-398

[4] Harrison MJ. Signaling in the arbuscular mycorrhizal symbiosis. Annual Review of Microbiology. 2005;**59**:19-42

[5] Gupta ML, Prasad A, Ram M, Kumar S. Effect of the vesicular-arbuscular mycorrhizal (vam) fungus *glomus fasciculatum* on the essential oil yield related characters and nutrient acquisition in the crops of different cultivars of menthol mint (*mentha arvensis*) under field conditions. Bioresource Technology. 2002;**81**:77-79

[6] Auge RM, Schekel KA, Wample RL. Leaf water and carbohydrate status of va mycorrhizal rose exposed to drought stress. Plant and Soil. 1987;**99**:291-302

[7] van Brussel AAN, Tak T, Boot KJM, Kijne JW. Autoregulation of root nodule formation: Signals of both symbiotic partners studied in a split-root system of *Vicia sativa* subsp. nigra. Molecular Plant-Microbe Interactions. 2002;**15**:341-349

[8] Takacs T, Cseresnyes I, Kovacs R, Paradi I, Kelemen B, Szili-Kovacs T, et al. Symbiotic effectivity of dual and tripartite associations on soybean (*Glycine max* L. *Merr.*) cultivars inoculated with bradyrhizobium japonicum and am fungi. Frontiers in Plant Science. 2018;**9**

[9] Abdel-Salam E, Alatar A, El-Sheikh MA. Inoculation with arbuscular mycorrhizal fungi alleviates harmful effects of drought stress on damask rose. Saudi Journal of Biological Sciences. 2018;**25**:1772-1780

[10] Chandrasekaran M, Chanratana M, Kim K, Seshadri S, Sa T. Impact of arbuscular mycorrhizal fungi on photosynthesis, water status, and gas exchange of plants under salt stress—A meta-analysis. Frontiers in Plant Science. 2019;**10**

[11] Grumberg BC, Urcelay C, Shroeder MA, Vargas-Gil S, Luna CM. The role of inoculum identity in drought stress mitigation by arbuscular mycorrhizal fungi in soybean. Biology and Fertility of Soils. 2015;**51**:1-10

[12] Goicoechea N, Merino S, Sanchez-Diaz M. Contribution of arbuscular mycorrhizal fungi (amf) to the adaptations exhibited by the deciduous shrub anthyllis cytisoides L. Under water deficit. Physiologia Plantarum. 2004;**122**:453-464

[13] Chappelle EW, Kim MS, McMurtrey JE. Ratio analysis of reflectance spectra (rars)—An algorithm for the remote estimation of the concentrations of chlorophyll-a, chlorophyll-b, and carotenoids in soybean leaves. Remote Sensing of Environment. 1992;**39**:239-247

[14] Blackburn GA. Quantifying chlorophylls and caroteniods at leaf and canopy scales: An evaluation of some hyperspectral approaches. Remote Sensing of Environment. 1998;**66**:273-285

[15] Zarco-Tejada PJ, Miller JR, Noland TL, Mohammed GH, Sampson PH. Scaling-up and model inversion methods with narrowband optical indices for chlorophyll content estimation in closed forest canopies with hyperspectral data. IEEE Transactions on Geoscience and Remote Sensing. 2001;**39**:1491-1507

[16] Penuelas J, Pinol J, Ogaya R, Filella I. Estimation of plant water concentration by the reflectance water index wi (r900/r970). International Journal of Remote Sensing. 1997;**18**: 2869-2875

[17] Hardisky M, Lemas V, Smart R. The influence of soil salinity, growth form, and leaf moisture on the spectral radiance of spartina alterniflora canopies. Photogrammetric Engineering and Remote Sensing. 1983;**49**:77-83

[18] Raymond Hunt E Jr, Rock BN. Detection of changes in leaf water content using near- and middle-infrared reflectances. Remote Sensing of Environment. 1989;**30**:43-54

[19] Yoder BJ, Pettigrew-Crosby RE. Predicting nitrogen and chlorophyll content and concentrations from reflectance spectra (400-2500 nm) at leaf and canopy scales. Remote Sensing of Environment. 1995;**53**:199-211

[20] Smith KL, Steven MD, Colls JJ. Use of hyperspectral derivative ratios in the red-edge region to identify plant stress responses to gas leaks. Remote Sensing of Environment. 2004;**92**:207-217

[21] Jiang JB, Chen YH, Huang WJ. Using hyperspectral remote sensing to estimate canopy chlorophyll density of wheat under yellow rust stress. Spectroscopy and Spectral Analysis. 2010;**30**:2243-2247

[22] Winterhalter L, Mistele B, Schmidhalter U. Assessing the vertical footprint of reflectance measurements to characterize nitrogen uptake and biomass distribution in maize canopies. Field Crops Research. 2012;**129**:14-20

[23] Tilling AK, O'Leary GJ, Ferwerda JG, Jones SD, Fitzgerald GJ, Rodriguez D, et al. Remote sensing of nitrogen and water stress in wheat. Field Crops Research. 2007;**104**:77-85

[24] Aliasgharzad N, Neyshabouri MR, Salimi G. Effects of arbuscular mycorrhizal fungi and bradyrhizobium japonicum on drought stress of soybean. Biologia. 2006;**61**: S324-S328

[25] Rapparini F, Peñuelas J. Mycorrhizal Fungi to Alleviate Drought Stress on Plant Growth. In: Miransari M, editors. Use of Microbes for the Alleviation of Soil Stresses. New York: Springer; 2014

[26] Giri B, Kapoor R, Mukerji KG. Improved tolerance of acacia nilotica to salt stress by arbuscular mycorrhiza, glomus fasciculatum may be partly related to elevated k/na ratios in root and shoot tissues. Microbial Ecology. 2007;**54**:753-760

[27] Gitelson AA, Zur Y, Chivkunova OB, Merzlyak MN. Assessing carotenoid content in plant leaves with reflectance spectroscopy. Photochemistry and Photobiology. 2002;**75**: 272-281

[28] Horler DNH, Dockray M, Barber J. The red edge of plant leaf reflectance. International Journal of Remote Sensing. 1983;**4**:273-288

[29] Feng W, Guo B-B, Wang Z-J, He L, Song X, Wang Y-H, et al. Measuring leaf nitrogen concentration in-winter wheat using double-peak spectral reflection remote sensing data. Field Crops Research. 2014;**159**:43-52

[30] Stagakis S, Markos N, Sykioti O, Kyparissis A. Monitoring canopy biophysical and biochemical parameters in ecosystem scale using satellite hyperspectral imagery: An application on a phlomis fruticosa mediterranean ecosystem using multiangular chris/proba observations. Remote Sensing of Environment. 2010;**114**:977-994

Chapter 7

Soybean (*Glycine max* (*L.*) *Merr.*) Production in the Cameroonian Cotton Basin between the Dynamics of Structuring an Agricultural Value Chain and Sustainability Issues

Eric Joël Fofiri Nzossié and Christophe Bring

Additional information is available at the end of the chapter

http://dx.doi.org/10.5772/intechopen.93981

Abstract

Over the past ten years, the Cameroonian cotton front, in the Sudano-Sahelian region, has experienced dynamics in soybean production, resulting in significant changes in agro-systems. From a simple hut culture not referenced in regional agricultural statistics, since 2010, soybean ranks 2nd in legumes cultivated after peanuts, followed by cowpea and voandzou, yet culturally and economically more adopted by the populations. The rapid development of cultivated areas from 6,705 ha in 2008 to 15,020 ha in 2018 is indicative of the enthusiasm shown by farmers for this speculation which, despite the almost lack of supervision on the part of public authorities, now structures a real value chain with challenges for improving farmers' incomes, in a space where more than 80% of working people live from agricultural activities. The objective of this research is therefore to analyse the sustainability challenges of a productive and market dynamics built around speculation subject to strong demands of international competitiveness. The results show that the viability of this dynamics remains precarious through a set of constraints inherent in the proven inability of farmers to empower themselves to optimise production and control marketing, as well as the poor technical supervision of cultivation.

Keywords: soybean, dynamic agriculture, cotton front, Sudano-Sahelian, Cameroon

IntechOpen

1. Introduction

Price volatility of agricultural raw materials on international markets (cotton, cocoa, coffee, etc.)[1] and its consequences on national food production, as well as the recurrence of food crises in many African countries over the past fifteen years, have repositioned agricultural issues on the agendas of governments of both developing and developed countries and international institutions. In addition to the commitments made by States at the 1996 World Food Summit, notably the increase of budget allocation for agriculture by at least 10%, and on the occasion of numerous world meetings (G8 Agriculture Summit of 2009, G20 Agriculture of 2011…), international and national responses to these two major global crises have mainly mobilised two converging agronomic trajectories: agricultural intensification (in the sense higher yields through better access to synthetic inputs and irrigation), and crop diversification to increase domestic food supplies and improve farmers' incomes.

Soybean is a good example of the realisation of this second trajectory in the Sudano-Sahelian zone. Its dynamics of spatial diffusion and production observed in ten years is an invitation to put into perspective many pessimistic, even alarmist judgements, on the capacity of African agriculture to meet the multiple challenges of increasing and diversifying agricultural production, improving farmers' incomes and supplying domestic agro-industries. These judgements are based in particular on still weak agricultural growth to alleviate rural poverty, the low representativeness of agriculture in international trade, the impact of price cuts and fluctuations of the main export products on the producers who are gradually integrated into logics of market economy and productivity [2]. In this author's opinion, nothing in these trends led to think that African agriculture is preparing to face economic, social and ecological challenges whose magnitude is still unprecedented in history, thus positioning agriculture in Sub-Saharan Africa, in a unique situation compared to other continents.

Experimentation, support for the dissemination and development of soybeans in the Sudano-Sahelian area in Cameroon constituted a strategic option for the Cotton Development Company (SODECOTON). The idea was to create added value in the face of volatile cotton prices on international markets. Diversification aimed at meeting the revitalisation needs of a society polarised by the cotton sector (support for farmers' contractual production, export of fibres and production of vegetable oil based on cotton seeds).

Thus, since 2008, this area, which shelters the cotton basin, is experiencing a dynamics of soybean production which involves profound changes in Sudano-Sahelian agro-systems and participates in the (re) configuration of new market relations between producers, traders, distribution intermediaries and agro-industrial companies. Available official statistics indicate that regional production increased from 41 tonnes in 2007 to almost 14,000 tonnes in 2017, while the total production in Cameroon for the same period was 7,801 and 18,886 tonnes, respectively.[2] This development followed the global trend marked by significant growth over

[1] For example, [1], evaluating the rates of change of certain raw materials, notes that with an average variability of 25.8% over the period 2003–2012 according to the instability index calculated by the United Nations for trade and development (UNCTAD), cotton prices are among the most volatile, far ahead of those of wheat (18.2%) or coffee (17.3%).
[2] FAOSTAT http://www.fao.org/faostat/fr/#data/QC accessed on June 23, 2020.

the past 25 years (123% between 1996 and 2004), particularly in Latin America [3, 4]. Since 2010, soybean ranks 2nd among the legumes cultivated in the cotton basin after peanuts, followed by cowpea and voandzou, which are, however, culturally and economically more adopted by the populations [5, 6].

Nearly 60% of cotton producers have either abandoned cotton growing, diversified food crops or replaced those less profitable and poorly supervised by public and private actors supporting agricultural development (case of *Vigna unguiculata L. walp - cowpea -*). The dominant trend that consists in establishing monoculture plots causes agro-systems reconfiguration and modifies rotations.

In an increasing number of family farms, cotton has thus ceased to be the head of crop rotation for the benefit of soybeans and incidentally peanuts.

Most of the regional production fuels partially estimated[3] national and cross-border agro-industrial demand for Nigeria. Macroeconomic data show that Cameroon imports an average of 20,000 tonnes of soybeans worth approximately CFAF 10 billion[4] a year, and GMO soybean meal worth CFAF 14 billion,[5] hence the challenge to increase domestic supply to meet agro-industrial demand. The rapid evolution of cultivated areas from 6,705 ha in 2008 to 15,020 ha in 2018 (FAOSTAT, Op.Cit.) is indicative of farmers' enthusiasm for this speculation which benefits little from the supervision of public authorities. Its rapid development in the Cameroonian cotton basin, where the income of more than 80% of agricultural assets historically depends on the sale of cotton, creates a need to understand the challenges of this development.

The above context raises the following research question: Can the productive and market dynamics on soybeans support the construction of a sustainable value chain to meet agro-industrial demand and improve farmers' incomes? The objective of this research paper is to analyse the sustainability challenges of a productive and market dynamics built around a speculation which is subject to strong demands of international competitiveness and offers a strong potential for improving farmers' incomes. The reflection initiated is based on the postulate that soybean brings out the structuring elements of a value chain, but its viability remains precarious by a set of constraints inherent in the proven inability of farmers to empower themselves to optimise production and control marketing, as well as poor technical supervision of culture. These constraints jeopardise the prospects for the sustainability of a soybean value chain despite the guarantee of a proven agro-industrial market.

1.1 State of knowledge on the topic

Soybeans (*Glycine max (L.) Merr.*) are today one of the main and important sources of proteins (40–42%) and vegetable oil (18–22%) used in human food [7, 8]. However, as with many other crops, its yield is limited in Sub-Saharan Africa by several factors, including poor cultivation practices, inadequate choice of land for its cultivation, degradation of soil fertility, climate

[3] Household consumption remains marginal, soybeans having so far integrated the eating habits of the Sudano-Sahelian populations compared to the southern part of the country.
[4] EUR 15, 244, 900
[5] EUR 2,342,900

change [9]. The orientation of this research highlights three major issues that have so far structured the production axes of knowledge on soybeans: those relating to socio-economic diversification in rural Africa, those relating to energy diversification and the fight against climate change, and finally those inherent in the polarisation of soybean productive dynamics by the agrifood industry in Latin America and Sub-Saharan Africa.

1.1.1 Soybeans in socio-economic diversification challenges in African rural areas

It was at the beginning of the 1970s that the IITA (International Institute of Tropical Agriculture) began its programme to improve African soybean lines, based on the crossing of high productivity lines of Asian origin and TGx lines (Tropical Glycine Cross). This work will, for example, enable Nigeria to boost the development of soybean production from 75,000 tonnes in 1980 (yield below 300 tonnes/ha) to 758,033 tonnes in 2018 (on average 1 tonne/ha) (FAOSTAT[6]). This speculation is now considered in Africa as one of the solutions that can improve the quality of food for populations, but also contribute to the diversification of farmers' incomes. Agronomic challenges for its establishment remain, however, significant; efforts aim at improving yield (1.2 tonne/ha on average), seed viability, fertilisation and inoculation of soybeans.

As early as the 1980s, soybean became part of the agro-systems of many countries in Sub-Saharan Africa to ensure self-sufficiency in foodstuffs, along with incentives. The lack of processing opportunities will lead farmers in many countries to abandon it. In Cameroon, its popularisation dates back to fifteen years, but domestic demand was quick to take off (agro-industries producing vegetable oil, infant flour and dairy products, animal feed). However, this dynamics is scientifically poorly documented apart from sparse statistics.

1.1.2 Soybeans in the global challenges of energy diversification and the fight against climate change

The past twenty years have been marked by an intense development of production and uses of biofuels in the world. From 1996 to 2006, the tonnage increased from around 16 million tonnes to almost 46 million [10]. This recent expansion has strengthened the relationship between agricultural (cereals, oilseeds and pulses) and energy markets, a situation favoured by national and regional policies. In 2019, the European Union (EU) authorised the import of American soybeans for the production of biofuel until 2021. Available statistics show that 75% of the EU soybean market share is covered by US production. However, it should be noted that interest in soybeans as a fuel source contributes to its growth in countries such as Argentina, whose production of soy-based agro-fuel exceeded that of Brazil in 2011. Agrifuels will therefore remain one of the engines of demand for soybeans, and analysts predict the probability of strong production growth by 2025 [11].

From an environmental point of view, soybeans are capable of capturing nitrogen from the air for their own growth and enrichment of their seeds with proteins, but also to fix it in the soil. Thus, its cultivation contributes to the fertilisation of the soil which would limit the prospects

[6] FAOSTAT (http://www.fao.org/faostat/fr/#data/QC), accessed on June 07, 2020.

of intensifying the use of chemical inputs and its corollary on the soils and water resources. Greenhouse gas emissions linked to seed production would be considerably reduced [12]. However, on the other hand, the extension of cultivated areas contributes significantly to deforestation and the degradation of plant cover [13, 14]. The impact of agricultural value chains on the environment increasingly underpins research dynamics, like the research initiative funded by the European Union through the "Value Chain Analysis For Development" Project (VCA4D) from 2016 to 2019, involving 24 countries in Sub-Saharan Africa, Central America and Southeast Asia.[7]

1.1.3 Food industry: A vector for accelerating the development of soybean farming

The food industry is one of the industrial sectors that have experienced particularly remarkable development over the past 20 years. This development has been accompanied by an extraordinary diversification of the supply of processed products based on cereals, legumes, oilseeds, etc. Global meat consumption statistics, for instance, show that almost three-quarters of the world's soybeans are used to feed animals, including poultry and pork. Between 1967 and 2007, pig meat production increased by 294%, that of eggs by 353% and that of poultry meat by 711%. Over the same period, the relative cost of these products has decreased. As the world's primary source of animal feed, soybeans have become an indispensable part of the intensive agricultural model [4]. Soybeans make up the Bulk of Dairy products, Oils and Flours. Food industry has become a driving force for production in Sub-Saharan Africa through a network of national companies which polarise the productive dynamics particularly oriented towards organic production. This third axis needs to be documented with regard to innovations on technical routes, marketing and transformation.

1.2 Theoretical framework of the study

The study mobilises the theory of spatial diffusion whose interest has been revived in contemporary geographic studies by [15, 16] with the contributions of [17–19]. These pioneering works emphasise the importance of temporality and spatiality in any diffusion process. Because any diffusion requires contacts between transmitters and receivers, these contacts implying contiguity or spatial connectivity [20, 21]. The theory is appropriate for studying the processes that involve the movement of materials, products, people, practices, or ideas together. It structures three main stages: the initial stage of appearance and progressive growth of the phenomenon observed, the intermediate stage of acceleration of growth, and the final stage of saturation and decay. In social sciences, the notion of diffusion is associated with innovation, from which it is inseparable. Introduction and popularisation trajectories of soybeans fit well with the dynamics of innovation according to Schumpeter (1934). This innovation supposes the combination of new things which propagate in an environment by causing irreversibilities in that environment's evolution (varietal selection, technical routes,

[7] Agrinature, *Value Chain Analysis For Development* » (VCA4D) https://europa.eu/capacity4dev/value-chain-analysis-for-development-vca4d- (accessed on June 23, 2020).

choice of soils …). The mobilisation of this theory will make it possible to decipher the trajectory of soybean diffusion in an attempt to grasp its sustainability.

The study uses the sector approach to complement the reality of the construction of the value chain analysed. Using [22] definition, a sector study is a very precise analysis of a whole system generated by a product. The concept refers to the economic analysis of a sequence of technically complementary physical operations enabling the creation, circulation and consumption of a good (or a service) [23, 24]. The supply chain approach constitutes an effective approach for structuring analysis in contexts of fragile institutional environments and failure of statistical information systems [25]. This fragility is inherent in the context of this study to the reliability of the data provided by the various contradictory sources of information. The mobilisation of this approach will make it possible to obtain knowledge on the functioning of the different segments that make up the sector, with a view to better deciphering the relationships between actors in these segments [26].

1.3 Adopted methodological approach

Data were collected using two approaches. The first focused on the statistical system on soybean production which combines the source of the Ministry of Agriculture through its branches at regional and divisional level on the one hand and that of FAO on the other hand. However, significant discrepancies between these two sources called for an arbitration which favoured data from the departmental branches closer to the field, despite the doubt about their reliability. This choice is supported by the monitoring of food production in the cotton area since 2006.[8]

The second approach was the subject of a diagnosis with actors of the sector through a questionnaire survey carried out between 2017 and 2019 among 500 farmers' members of the soybean producer groups chosen from the snowball technique. Questionnaires were directly administered to farmers in Touboro, Madingring, Mayo-Rey, Tchollíré sub-divisions, on the motivations for growing soybeans, marketing of the production, difficulties encountered, various types of support. Interviews were conducted with officials of the Soybeans Processing Industry of Cameroon (SOPROICAM), an agrifood company that governs the soybean sector in the Sudano-Sahelian zone, officials of the Cameroonian dairy company (Camlait) and service personnel of the Ministry of Agriculture premises.

Survey data were subjected to statistical processing in Excel to identify trends in production and marketing. Interviews were the subject of a qualitative treatment using the content analysis method.

In a first articulation, results present conditions and modalities of soybean farming in Cameroonian cotton basin. The second articulation describes the methods of building a value chain around speculation, and whose sustainability is questioned in a third articulation based on the difficulties and constraints raised by the stakeholders interviewed.

[8] Monitoring the development of food production in the Cameroonian cotton zone was the subject of a regional research project involving Cameroon, Chad and the Central African Republic from 2004 to 2009 [27].

http://dx.doi.org/10.5772/intechopen.93981

2. Terms and conditions for developing soybean cultivation in the Cameroonian cotton basin

2.1 Cotton crisis as a vector to popularise soybeans

Cotton cultivation is at the heart of multiple challenges in the Sudano-Sahelian zone of Cameroon (**Figure 1**) [28–30]. The sector experienced a serious crisis from 2006 to 2011 following the constitution by China of large stocks of cotton fibre (EU-ACP, 2015), with the consequence of a drastic fall in production from 315,000 tonnes in 2005 to 180,000 tonnes in 2011 (FAOSTAT). The number of producers increased from 360,000 in 2006 to around 250,000 over the same period, following the drop in the purchase price of seed cotton from producers. The consequences of this fall in the purchase price were dramatic for the producer, given the function of supporting all the charges for agricultural inputs of food crops (cereals, legumes) that cotton plays [28, 29, 31]. They were more so for SODECOTON (drastic fall in the tonnage of exported fibre and seeds for the production of vegetable oil and cottonseed meal for animal feed).

In 2006, SODECOTON initiated a crop diversification project through a feasibility study of soybeans and sunflower as a rotation crop with cotton [32]. This initiative will be avant-garde to the most severe production crisis that SODECOTON will experience between 2008 and 2011, with respectively 185,000 tonnes of unginned cotton produced, 140,000 tonnes, 190,000 tonnes and 180,000 tonnes, compared to 315,000 tonnes in 2005 (peak production which has never reached SODECOTON again[9]). The results of the study foreshadowed the prospects for a possible diversification crop for soybeans,[10] mainly in the cotton front (Mayo-Rey Division) unlike sunflowers.

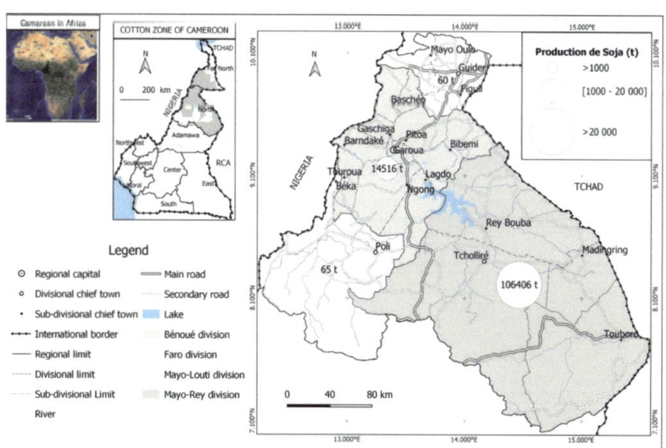

Figure 1. The Cameroonian cotton basin, a soybean production area.

[9] Data available on the FAO website (http://www.fao.org/faostat/fr/#data/QC), accessed on June 28, 2020.
[10] The so-called "local" soybean varieties (Houla 1, Houla 2, TGX-849-29-4D, Carrefour Nari, Ngong and Pitoa 2) had an average yield of 1600 kg / ha, more attractive than the varieties imported from IITA of Ibadan and Brazil [31].

The SODECOTON diversification project was implemented as from the 2008 agricultural season through two strategic pillars: the pre-financing of agricultural inputs (fertilizers, herbicides) intended for soybeans for cotton producers[11]; and the purchase of crops. The rapid increase in production (**Figure 2**) however, was followed by four difficulties which led to the termination of the project after 2011:

- Persistence in 2011 of the cotton crisis and its consequences on the financial resources of SODECOTON leading to the end of the pre-financing of inputs and the purchase of soybean crops;
- Competition from Nigerian traders with producers to the detriment of SODECOTON, resulting in a low level of recovery of pre-financing of inputs;
- Increasing immobilisation of unsold stocks of soy products (oil cake, vegetable oil), prices being considered higher than those of cotton seed products on domestic markets;
- Technological difficulties encountered in processing more rustic soybeans than cotton seeds.

The cessation of the crop diversification project as a consequence of the above difficulties will stimulate a new productive and market dynamic supported by Nigerian wholesalers whose prices on the domestic markets are twice higher than those offered by SODECOTON and Soyabeans Processing Industrial of Cameroon (SOPROICAM), a cooperative company engaged in the purchase and industrial processing of soybeans, to conquer the regional market. The sale of crops on domestic markets becomes a direct source of income for producers, in an environment marked by permanent cash flow tensions.

Thus, between 2011 and 2016, cotton ceased to be the head of rotation in crop rotations in favour of soybeans for 40% of producers interviewed. This proportion was 60% after 2017, with a strong trend towards the proliferation of monoculture farms by clearing new plots.

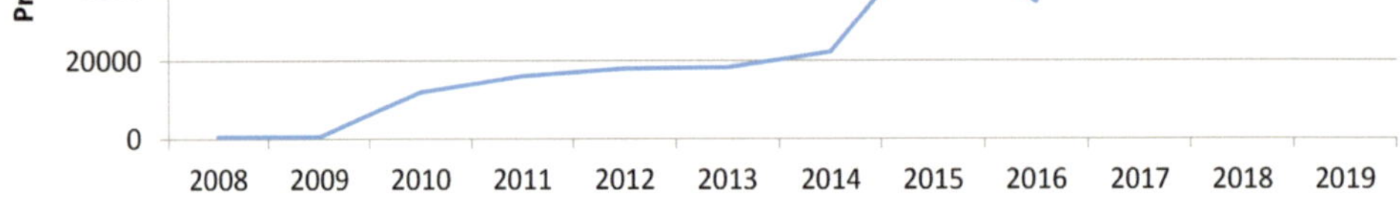

Figure 2. Evolution of soybean production in the cotton basin of Mayo-Rey (2008–2019).

[11] SODECOTON has for twenty years adopted the incentive policy for cotton production by providing credit inputs intended for the production of food crops to cotton producers only. This policy will simply be accentuated on soybeans as part of the crop diversification project.

http://dx.doi.org/10.5772/intechopen.93981

2.2 Towards a dynamics of extension of the production basin

In the North Region, the main production basin of the entire Sudano-Sahelian region, the cultivated area increased from less than 6,000 ha in 2008 to 68,000 ha in 2019.[12] The activity brings together nearly 50,000 producers, at least 35% of whom have abandoned cotton growing in favour of soybeans. And owing to their status as crops which are almost entirely intended for the domestic and cross-border market, the two speculations compete as much in the use of space as in family and salaried agricultural labour.

From a spatial point of view, soybeans are indeed developing in the new cotton basin of the Mayo-Rey Division, opened within the framework of the "Programme of migration and agricultural support services" implemented by the State of Cameroon from 1974 to 1997. This programme made it possible to move and supervise the settlement of more than 200,000 migrants from the Far North Region to areas with high agricultural potential in the North Region [33] (**Figure 2**). The two speculations are therefore cultivated by so-called "non-native" populations for whom access to land remains a challenge because of the socio-cultural barriers which limit their access to land ownership in settlements. Local populations have always lived their presence as a form of spoliation of their land [34]. In this context of precarious land, the average cultivated area per producer is small (1 ha), with however at least 50% of farms between ¼ and ¾ ha. This proportion is justified by the fact that soybeans are almost entirely cultivated by family farms (FF) which also ensure the production of foodstuffs intended primarily for self-consumption (millet/sorghum, corn, cowpeas, groundnuts). Thus, with an average yield of 1 tonne/ha, EFA production oscillates between 500 and 800 kg, or 5 to 8 bags of 100 kg. It is therefore essentially an atomised production on small areas exploited for an average duration of 5 years from the clearing. In some cases, soybeans are grown in rotation with corn one to two years to capitalise on residual effects of nitrogen. Despite this situation, which translates into a context of land insecurity for most producers, soybean monoculture farms are, nevertheless, developed on 5–15 ha, despite their low mechanisation. However, the overall increase in areas by clearing and deforestation will soon come up against the areas set aside for environmental protection (Bénoué and Bouba-Ndjidda national parks; Areas of Hunting Interest which cover nearly 50% of the North Region).

From a labour point of view, almost essentially manual production justifies the use of family labour force adapted to the crumbling context of plots on the one hand, and hired labour for farms greater than or equal to 2 ha on the other hand. According to technical soybean routes, the labour force is called upon to perform several tasks (**Table 1**):

- Preparing fields (clearing and/or cutting down of trees for the creation of new plots, burning);
- Sowing;
- Crops maintenance (fertiliser and herbicide spreading, weeding);
- Harvests;
- Threshing soybeans for seeds extraction.

---

[12] Source: North Regional Delegation for Agriculture and Rural Development (2020).

	Jan. Feb.	Mar. Apr May	June	July	Aug. Sept.	Oct. Nov.	Dec.
Indicative period	Until March	As from the 10th of March	As from the 15th	As from the 15th	During the two months	From the 15th to November	Until March
Activities	Continuous marketing	Land identification/field preparation	Early sowing	Late sowing	Crop maintenance	Start of the harvest	Threshing and marketing

Table 1. Distribution of tasks according to the soybean agricultural calendar.

The spread of labour-intensive tasks over ten months in the year (March to December) has thus contributed to the development of a subsector of agricultural workers who have become essential in the productive dynamics, remuneration being paid by task. For agricultural workers, this is an immediate source of income that is beginning to structure a seasonal migratory flow to soybean production areas. In addition, the growing use of labour force, whether family or salaried, is increasingly a factor in the variation/reduction of areas devoted to other food crops (millet/sorghum, maize, cowpeas, peanuts, cotton) due to the overlapping of agricultural calendars which puts the execution of agricultural tasks in competition. Cotton is more demanding in terms of labour force, and conversely, does not provide immediate income, the producer being just as dependent on uncertain payment terms. Overall, this growing craze for soybean cultivation underpins the consolidation of the producers' strategic link in the construction of a value chain.

3. Dynamics of building a soybean value chain

Michael Porter (1986) who is the author of the value chain concept developed it in his reference book "Competitive Advantage". The value chain describes all the activities necessary to carry out a product or a service from its conception, through different phases of production to its distribution to final consumers, then to its destruction after use. It breaks down activity into sequences of elementary operations and identifies sources of potential competitive advantage. This breakdown improves the overall performance of the value chain by strengthening each link and links between links.

As production is only one of the stages in creating added value, knowing the dynamics of building soybean value chain involves breaking down all of the activities that structure it, all linked to each other.

3.1 Tentative assessment of financial effects induced by the marketing of regional soybean production

Regional soybean production follows two marketing trajectories: contract buying and free selling on domestic markets. The first trajectory is supported by the Soybeans Processing Industry of Cameroon (SOPROICAM) which has set up a network of 20,000 farmers present in more than 2000 producer organisations. The contractual relationship is framed by two provisions guaranteeing the supply to producers of inputs (fertilisers and herbicides) at the start of the crop year and the obligation to deliver the crops to SOPROICAM at a price fixed in advance by the two parties. The kg fixed for the 2018–2019 agricultural campaign was CFAF 170 (or EUR 0.259), against CFAF 120–150 (EUR 0.182–0.228) on the domestic market. For this campaign, this company collected 6,000 tonnes of soybeans (60,000,100 kg bags) worth CFAF 1,020,000,000 (EUR 1,554,980). The unsatisfied purchase forecast for this campaign was 20,000 tonnes, which would have generated in the production segment alone a money supply in circulation in the region of CFAF 3.4 billion (EUR 5,183,270).

The second trajectory is organised around a network of wholesalers who source from producers in rural markets. These wholesalers act as intermediaries for cross-border or domestic agrifood companies. For example, the Cameroonian dairy company (Camlait), an agrifood company based in Douala, introduced, in 2019, 581,203 kg soybeans (5812 bags of 100 kg) in the manufacturing process of its milk-based products (Nouriss, Breaksoy and Riverr). This quantity represents between CFAF 69,744,360 and CFAF 87,180,450[13] depending on the price in rural markets. In addition to food products, soybeans are one of the main feedstocks for poultry feed. Domestic demand, however, remains unevaluated, as does that of Nigerian agrifood companies, which have supplied Nigerian wholesalers with an important soybean purchasing channel for the past decade.

The attempt to assess the financial effects generated from proven national demand reveals the existence of significant economic potential, the optimisation of which would contribute to improving the incomes of farmers directly involved in trade relations with other links of the value chain. Despite uncertainties in the regional production data noted in the methodological framework, reaching 100,000 tonnes of production since the 2018 agricultural campaign would thus represent a minimum annual monetary value of CFAF 12.5 billion (EUR 19,056,100) for producers at the average price on the rural market. This estimate does not include the production of isolated basins in the Far North Region, which is not taken into account in this study. It does not equally include the added value generated by other links in the chain (distribution, transformation).

3.2 Structuring elements of a soybean value chain

The marketing trajectories analysed above highlight convincing elements in the structuring of a value chain that links four links (supply of basic goods and services, production, distribution, transformation) (**Figure 3**).

The first link is built around four uncoordinated actions which nevertheless make it possible to effectively support the supply of basic agricultural services (seeds, fertilisers, herbicides and financing). Within the framework of its contractual relationship with producer groups, SOPROICAM ensures the supply of inputs to groups that express demand. These are mineral fertilisers of ternary combination N (nitrogen) - P (phosphate) - K (potassium) of formulation 12-24-12 and 20-10-10, as well as herbicides (glyphosate, roundup). The first formulation is specific to the production of legumes whose natural nitrogen production potential directs mineral requirements more towards phosphate. The second formulation which is more suited to the cultivation of maize, responds to a specific request from farmers to the company to support cereal production intended for self-consumption. This company also plays the role of guarantor with micro finance establishments for the granting of loans to producers.

Farmers' growing interest in soybeans has, however, led to the development of the chemical input marketing activity through a network of small distributors who supply rural markets. However, a weak segment of this link remains the supply of seeds, most of which are harvested. Since the introduction and popularisation of varieties adapted to ecology by SODECOTON from 2007 to 2010 (so-called "local" varieties: Houla 1, Houla 2, TGX-849-29-4D, ESA, SJ 235; varieties from the IITA collection of Ibadan in Nigeria: TGX 1448 2E, TGX 1485 1D and TGX 1910 14F), seed production has no longer been the subject of a specific

[13] Or EUR 106,325 and EUR 132,906.

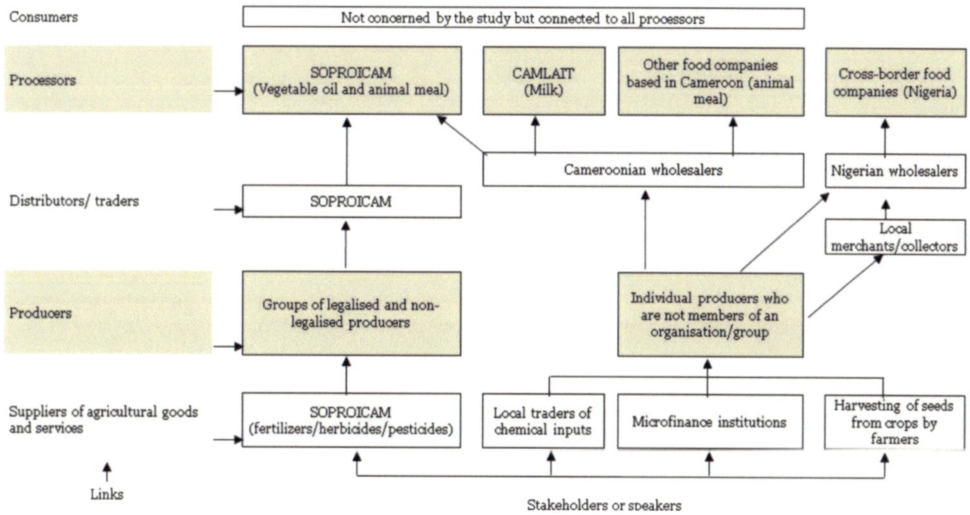

Figure 3. Graphical representation of the soybean value chain in the Sudano-Sahelian region of Cameroon.

research or experimentation programme. Consequently, the seeds in circulation come from the repeated reuse of seeds taken from the crops. Production is therefore mainly ensured from so-called "all-coming" varieties resulting from these re-uses. This context is partly maintained by the absence of specific requests on varieties by food companies. Colour therefore is the only criterion for differentiating soybeans between red and white.

The second link in the value chain brings together two categories of producers. The first brought together nearly 20,000 producers in groups whether they have a legal existence or not with the administration in charge of cooperative companies. The second concerns producers who are not involved in any form of collective dynamics. The first initiatives to organise soybean producers emerged from 2011 in the Mayo-Rey Division and will quickly evolve towards the creation of increasingly structured groups despite the low legalisation rate observed (less than 10%). As noted above, SOPROICAM maintains a network of more than 2000 groups. The file of cooperative societies in the North Region references ten legalised cooperatives of soybean producers since 2017, including six in the only Mayo-Rey Division which provides 80% of the production in the North Region (**Table 2**).

Women represent 20% of the group members. This small proportion could be explained by the density of labour-intensive tasks on the one hand, and the immediate financial profitability that the sale of soybeans provides arousing great interest among men on the other hand.

The third link represents the heart of the value chain. It is made up of SOPROICAM and a network of Cameroonian and Nigerian wholesalers, two actors in the governance of the soybean sector. Their respective role in marketing has already been explained in the functioning of the trajectories presented above (2.1.). If it is true that the first actor has invested in the control of the sector from the field to the factory, wholesalers meanwhile are agents of execution of the strategic choices of agrifood companies expressed through the formulation of quantitative and qualitative needs.

Sub-division	Number of cooperatives	Number of members
Rey-Bouba	1	700
Tcholliré	1	1000
Madingring	1	650
Touboro	3	2000

Source: Surveys of producers' cooperatives and base of cooperative societies of the North Regional Delegation for Agriculture and Rural Development (April 2020).

Table 2. Spatial distribution of cooperative producers of soybeans in Mayo-Rey (2019).

No.	Denomination	Products developed	Geographical location
1	SOPROICAM	- Vegetable oil - Oil cakes	Douala
2	CAMLAIT	- Nourished, - Breaksoy - Riverr	Douala
3	NT Foods S.A.R.L (Tanty, brand name)	- Soybeans flour - Reine custard powder infant soybeans porridge - Tanty custard powder	Yaoundé
4	Nkam Feed Company	Provende (poultry feed)	Yaoundé
5	Cameroon Provender Company	Provende (poultry feed)	Bafoussam

Source: authors' compilation.

Table 3. Main agri-food companies processing soybeans in Cameroon.

The fourth link brings together all of the processing companies whose activities relate to the production of vegetable oil, dairy products, edible flours, oil cakes and animal feed. The study identified five main companies under Cameroonian law, soybeans being one of the main raw materials in the production process (**Table 3**).

The existence of the above four links and the market relationships between them confirm the presence of structural elements in a soybean value chain. The growing investment of distribution and processing players in supporting the productive dynamics testifies today's challenges for a need to perpetuate regional supply to meet increasing domestic and cross-border demand.

4. What prospects for the sustainability of the soybean value chain in the Sudano-Sahelian zone in Cameroon?

The growing statistics on soybean production are likely to obscure the constraints which constitute potential vectors of weakening the productive dynamics noted above. Indeed, the diagnosis of the organisation and functioning of production as a strategic link in the value chain highlights three major constraints relating to the structuring of producers, the provision of agricultural services and the functioning of market.

With regard to the first constraint, the collaboration model implemented by SOPROICAM tends to polarise all producers towards the creation of structured and legalised groups. Despite the current proliferation of independent wholesalers, consolidating the distribution link means moving towards the establishment of true merchant cooperatives. This future trend implies for producers to build real organisations whose legal existence constitutes a foundation of credibility vis-à-vis other actors in the value chain. However, the low proportion of registration by the competent administration of existing structured groups with a view to obtaining the status of cooperatives results in difficulties in opening accounts in financial and banking establishments due to the lack of a reliable guarantee; refusal to pay membership fees by members; leadership struggles and internal confidence crises.

From the point of view of the provision of agricultural services, the support of distribution players to producers constitutes a dusting which, far from boosting production, keeps farmers in a spiral of dependence on buyers. Several farmers denounce the low level of inputs supplied by SOPROICAM (less than 40–50% of the needs expressed) in exchange for the exclusive purchase of the crops. The fixing of the purchase price at the start of the crop year thus reinforces this dependence of farmers on the company. This context therefore contributes to the establishment of an unsustainable debt cycle in the medium and long term by farmers, many of whom opt for direct sales on domestic markets.

This second constraint is also marked by the absence of a seed sector to sustainably increase yield. With an average yield of 1 tonne/ha, soybean productivity does not bode well for the prospects of sustainable growth, thus justifying the intensification of land clearing and deforestation to expand the cultivated areas, despite the virtual absence of mechanisation of production and technical supervision by State services.

As a corollary to the above, the regional soybean market is characterised by inter-annual price instabilities that are highly detrimental to farmers in terms of making production forecasts due to market uncertainty. In 2012 for example, the kg of soybeans cost CFAF 220 (EUR 0.335) against CFAF 120–150 (EUR 0.182–0.228) between 2017 and 2019. The strong presence on the domestic markets of Nigerian wholesalers frequently subject to variations in the cost of Naïra (devaluation) is a factor which reinforces market uncertainties.

Ultimately, beyond the institutional will displayed to support the soybean sector by the launch in 2014 of a dedicated agropolis in the Far North Region, it is difficult to see concrete actions on the ground, strengths of the State for the benefit of key players.

5. Conclusion

In this study, we have proposed to analyse the sustainability issues of the productive and market dynamics built around soybeans, speculation subject to strong demands for international competitiveness and which at the same time has a strong potential for improving farmers' incomes. The conditions and modalities for the development of this speculation corroborate the importance of temporality and spatiality in the soybean diffusion process, which is driven by SODECOTON in the face of the persistent context of the cotton crisis. The productive dynamics described responds to the spatial diffusion trajectory structured in three stages: the initial stage of appearance and growth of soybean cultivation from 2008, the intermediate stage of acceleration of growth observed since the 2015–2016 agricultural season, and the final stage of saturation and decay not yet started. The temporality of the first two stages testifies to the rapidity of the environment penetration through technical innovation which combined varietal selection, technical routes, choice of soils, trading strategy. This speed also shows the capacity of Sudano-Sahelian farmers to take ownership of innovations and optimise their value.

However, from an agronomic point of view, the low productivity (on average 1 tonne/ha) is evidence of the weight of the constraints which limit the optimisation of the productive potential of almost 2 tonnes/ha revealed by experimental trials in 2006, mainly on local varieties. The study highlights in particular the virtual absence of mechanisation of production, a situation affecting both medium and large farms from a spatial point of view, as well as the absence of a seed industry with the corollary of the use of so-called "all-comers" seeds. The major environmental issues in the Sudano-Sahelian region call for priority to be given to the development of intensification trajectories for soybean cultivation.

The growing development of production, although backed by a spatial sprawl trajectory, offers the opportunity to build a value chain that articulates four strategic links: provision of agricultural services, production, distribution and processing. As pointed out above, the investment of distribution and processing players in supporting the productive dynamics testifies to the challenges that the need to perpetuate regional supply raises today. The mobilisation of the value chain approach will have made it possible to grasp the relationships existing between the categories of actors that make up the value chain and to identify their weaknesses. The attempt to assess the financial effects induced by the marketing of regional production has above all revealed the potential that soybean speculation abounds in terms of the circulation of annual money supply.

Ultimately, the results obtained corroborate the hypothesis which maintained that soybean brings out the structuring elements of a value chain, but its viability remains precarious by constraints inherent in the proven inability of farmers to empower themselves to optimise production and control the marketing and poor technical supervision of the crop. It appears that these constraints are jeopardising the prospects for the sustainability of a soybean value chain despite growing demand. The control of all the activities of the chain by private actors constitutes a risk of reinforcement of dependence of farmers who are strongly subjected to the strategic choices of these actors. This observation suggests that public authorities should be strongly challenged in their regulatory role, but above all to support local dynamics by facilitating some procedures, such as streamlining the procedures for creating accounts to increase the proportion of cooperatives registered with the administration in charge of agriculture. The seed sector is a strategic segment which is almost neglected. The sustainability challenges of the soybean value chain are therefore strongly questioned.

Acknowledgements

The authors thank the services of the North Regional Delegation for Agriculture and Rural Development for facilitating the gathering of information.

Conflict of interest

The authors declare no conflict of interest.

Author details

Eric Joël Fofiri Nzossié[1,2,3]* and Christophe Bring[1,3]

*Address all correspondence to: fofiri_eric@yahoo.fr

1 Department of Geography, University of Ngaoundéré, Cameroon

2 Associate Researcher, UMR 201 "Development & Societies" IRD-IEDES, France

3 The Ministry of Environment, Protection of Nature and Sustainable Development, P.O. Box 320, Yaoundé, Cameroon

References

[1] Diasso Yankou, « Dynamique du prix international du coton: aléas, aversion au risque et chaos », *Recherches économiques de Louvain*, 2014/4 (Vol. 80), p. 53-86. DOI: 10.3917/rel.804.0053. URL: https://www.cairn.info/revue-recherches-economiques-de-louvain-2014-4-page-53.htm

[2] Griffon M. (2003). 'Quand l'agriculture africaine va-t-elle commencer à répondre aux enjeux du futur ?' (Griffon M. 2003). In *Cahiers Agricultures*, Volume 12, pp. 141-143.

[3] Arvor Damien. Etude par télédétection de la dynamique du soja et de l'impact des précipitations sur les productions au Mato Grosso (Brésil). Sciences de l'Homme et Société. Université Rennes 2, 2009. Français. tel-00422109v2

[4] WWF. 2014. The growth of soy: Impacts and solutions. WWF International, Gland, Suisse.

[5] Nyahnone T J., 2017. Le développement de la culture du soja dans les fronts d'extensions cotonniers du Nord-Cameroun: Le cas du Mayo-Rey. Mémoire de master de géographie, Université de Ngaoundéré, 135p.

[6] Nitcheu Wakponou Darryl Neil (2018). *Les enjeux environnementaux du développement de la chaîne de valeur soja dans le bassin cotonnier du Nord-Cameroun: cas des arrondissements de Touboro et de Madingring,* Mémoire de master de géographie, Université de Ngaoundéré, 165p

[7] Mehmet Oz., 2008. Nitrogen rate and plant population effects on yield and yield components in soybean. *African Journal of Biotechnology* Vol. 7 (24), pp. 4464-4470, 17 December, 2008, pp 4464-4470.

[8] Baboy L., Kidinda L., Kilumba M., Langanu S., Mazinga M., Tshipama D., Kimuni L., 2015. Influence du semis tardif sur la croissance et le rendement du soja (*Glycine max*) cultivé sous différents écartements à Lubumbashi. RD Congo. *International Journal of Innovation and Applied Studies,* pp 104-109.

[9] Kananji G., Yohane E., Siyeni D., Kachulu L., Mtambo L., Chisama B., Mulekano O., 2013. A Guide to Soybean Production in Malawi. Department of Agricultural Research Services (DARS), Lilongwe, Malawi. Retrieved November 29

[10] Dronne Y., Forslund A., Gohin A., Guyomard H., Levert F. (2007). 'Impacts du développement des biocarburants aux États-Unis et dans l'UE sur les marchés internationaux de produits de grandes cultures'. In *Oilseeds and fats, Crops and Lipids (OCL),* VOL. 14 N° 6 novembre-décembre 2007, pp. 347-353 (https://www.ocl-journal.org/articles/ocl/pdf/2007/06/ocl2007146p347.pdf).

[11] Hart Energy, 2013. Global biofuels outlook to 2025. Global biofuels center (https://ubrabio.com.br/sites/1800/1891/PDFs/InformacaodoSetor/2013BiodieselPrimer.pdf).

[12] Schneider A., Huyghe Christian (Coord.), 2015. *Les légumineuses pour des systems agricoles et alimentaires durables,* Versailles, Editions Quae, Collection Technology & Engineering - 512 p. (https://books.google.cm/books?id=OyQ0CgAAQBAJ&pg=PA381&lpg=PA381&dq=Sojaxa,+2013&source=bl&ots=Ia8f0L7cn3&sig=ACfU3U3JH8Dwf9_tteL5lsnHlkZokekE9g&hl=fr&sa=X&ved=2ahUKEwivw4zA5MHqAhUL3BoKHXT_CvIQ6AEwEnoECAcQAQ#v=onepage&q=Sojaxa%2C%202013&f=false)

[13] Le Tourneau François-Michel (2004). Jusqu'au bout de la forêt ? Causes et mécanismes de la déforestation en Amazonie brésilienne. In, Mappe Monde (ISSN: 0764-3470), Montpellier, Published by Maison de la géographie, numéro 75, pp. 1-25.

[14] Guéneau Stéphane, « Durabilité des chaînes globales de valeur du soja et de la viande de bœuf en Amazonie: conséquences d'une gouvernementalité néolibérale », Brésil(s) [En ligne], 13 | 2018, mis en ligne le 31 mai 2018, consulté le 23 juin 2020. URL: http://journals.openedition.org/bresils/2468 ; DOI: https://doi.org/10.4000/bresils.2468

[15] Hàgerstrand T. (1952). « The propagation of innovation waves ». Lund studies in geography, ser. B, n° 4.

[16] Hàgerstrand T. (1957). « Migration and area ». Lund studies in geography, sery. B, n° 13.

[17] Saint-Julien Th. (1985). La diffusion spatiale des innovations. Montpellier GIP RECLUS, coll. « RECLUS Modes d'Emploi ».

[18] Saint-Julien Th. (1992). « Diffusion spatiale ». Encyclopédie de géographie. Paris Economica.

[19] Gould P., DiBiase D., Kabel J. (1990). « Le SIDA la carte animée comme rhétorique cartographique animée ». Mappemonde, n° 1.

[20] Dumolard Pierre. Accessibilité et diffusion spatiale. In: Espace géographique, tome 28, n°3, 1999. pp. 205-214; doi: https://doi.org/10.3406/spgeo.1999.1254 https://www.persee.fr/doc/spgeo_0046-2497_1999_num_28_3_1254

[21] Maudet Jean-Baptiste, « La diffusion spatiale à l'épreuve des rodéos. Les hybridations tauromachiques de l'Europe à l'Amérique », *Ethnologie française*, 2011/4 (Vol. 41), p. 667-675. DOI: 10.3917/ethn.114.0667. URL: https://www.cairn.info/revue-ethnologie-francaise-2011-4-page-667.htm

[22] Terpend N. (1997). *Guide pratique de l'approche filière. Le cas de l'approvisionnement et de la distribution des produits alimentaires dans les villes*, DT/18-97F, FAO, Collection « Aliments dans les villes », 34 p.

[23] Labonne M. (1987). 'Sur le concept de la filière en économie agro-alimentaire'. In Kermel-Torrès Doryane (ed.), Roca P.J. (ed.), Bruneau Michel (ed.), Courade Georges (éds). *Terres, comptoirs et silos: des systèmes de production aux politiques alimentaires*, Actes du Colloque « Séminaire interdisciplinaire sur les Politiques Alimentaires », Paris, ORSTOM, pp. 137-149.

[24] Aragrande Maurizio (1997). *Les approches disciplinaires de l'analyse des* SADA, Communication présentée au Séminaire sous-régional FAO-ISRA « Approvisionnement et distribution alimentaires des villes de l'Afrique francophone », Dakar, 14-17 avril 1997, Collection « Aliments dans les villes », 55 p.

[25] Temple L., Lançon F., Montaigne E., Soufflet J-F. (2009a). 'Introduction aux concepts et méthodes d'analyse de filières agricoles et agro-industrielles'. In *Economies et Sociétés*, Série « Systèmes agro-alimentaires », AG, n°31, 11/2009, pp. (1803-1812).

[26] Scott G. et Griffon D. (1998). *Méthodes pour analyser la commercialisation agricole dans les pays en développement*, Montpellier, CIRAD, Paris, Karthala, Lima, CIP/Apartodo, 495 p

[27] Ndjouenkeu R., Fofiri Nzossie E.J., Kouebou C. P., Njomaha C., Koussou Miand O., Grembombo A. 2010. *Le maïs et le niébé dans la sécurité alimentaire urbaine des savanes d'Afrique centrale*, Communication présentée au Symposium Innovation & Sustainable Development in Agriculture and Food, CIRAD-INRA-Montpellier SupAgro (28 juin – 01 juillet 2010, Montpellier, France)

[28] Devèze Jean-Claude, « Le coton, moteur du développement et facteur de stabilité du Cameroun du Nord ? », *Afrique contemporaine*, 2006/1 (n° 217), p. 107-120. DOI: 10.3917/afco.217.0107. URL: https://www.cairn.info/revue-afrique-contemporaine-2006-1-page-107.htm

[29] Levrat R., 2009. Le coton dans la zone franc depuis 1950. Un succès remis en cause, Paris, Editions L'Harmattan, Collection Etudes africaines, 248 p.

[30] Folefack Denis Pompidou. Coordination des acteurs dans un contexte de crise: le cas de la filière coton au Cameroun depuis 1990. Economies and finances. Université

Rennes 2; Université Européenne de Bretagne, 2010. French. <NNT: 2010REN20030>. <tel-00597413>

[31] Folefack et NTSOU BAKWOWI Jeshma, Cokou KPADE Patrice. J. Appl. Biosci. 2014. La crise de la filière cotonnière et sécurité alimentaire au Nord Cameroun, in: Journal of Applied Biosciences 75:6221-6231

[32] Wey J et Ibrahima S., 2006. *Cultures de diversification: Étude de faisabilité du soja et du tournesol dans la zone cotonnière du Nord Cameroun,* Rapport CIRAD, 55 p. (http://agritrop.cirad.fr/559886/1/document_559886.pdf)

[33] Fofiri Nzossie E.J., 2013. *Les déterminants de l'offre alimentaire vivrière dans les villes du Nord-Cameroun,* Thèse de Doctorat de géographie économique, Université de Ngaoundéré, Cameroun, 441 p. (https://tel.archives-ouvertes.fr/tel-01092376)

[34] Seignobos C. (2006). *'Colloque international SFER "Les frontières de la question foncière",* Montpellier, France, 2006, 21 p.